Happiness Studies Book Series

For further volumes:
http://www.springer.com/series/10117

Happiness Studies Book Series

Series Editor:

Aims and Scope

Exploring features and implications for personal and social empowerment from a substantially interdisciplinary point of view.
Like the Journal of Happiness Studies, the series explores happiness through both objective and subjective indicators. Subjective aspects comprise cognitive evaluations (like life satisfaction), positive affect and emotions, development of meanings, competences, and goals.
Key issues includes appraisal of life, work conditions, mental and physical health, developmental trajectories throughout the life span, socio-economic conditions, cultural aspects, and their impact on individual and social wellbeing.

Hilke Brockmann · Jan Delhey
Editors

Human Happiness and the Pursuit of Maximization

Is More Always Better?

Springer

Editors
Hilke Brockmann
Jan Delhey
School of Humanities and Social Sciences
Jacobs University Bremen
Bremen
Germany

ISSN 2213-7513 ISSN 2213-7521 (electronic)
ISBN 978-94-007-6608-2 ISBN 978-94-007-6609-9 (eBook)
DOI 10.1007/978-94-007-6609-9
Springer Dordrecht Heidelberg New York London

Library of Congress Control Number: 2013941346

Printed on acid-free paper

Springer is part of Springer Science+Business Media (www.springer.com)

Contents

Contributors

Kurt Bayertz Philosophical Seminar, University of Münster, Münster, Germany, e-mail: bayertz@uni-muenster.de

Kent C. Berridge Affective Neuroscience & Biopsychology Lab, Department of Psychology, University of Michigan, Ann Arbor, MI, USA, e-mail: berridge@umich.edu

Mathias Binswanger Institute for Competitiveness and Communication, School of Business of the University of Applied Sciences Northwestern Switzerland, Olten, Switzerland, e-mail: mathias.binswanger@fhnw.ch

Hilke Brockmann School of Humanities and Social Sciences, Jacobs University Bremen, Bremen, Germany, e-mail: h.brockmann@jacobs-university.de

Robert Davidson Amsterdam Research Center for Gender and Sexuality, University of Amsterdam, Amsterdam, The Netherlands, e-mail: R.J.Davidson@uva.nl

Jan Delhey School of Humanities and Social Sciences, Jacobs University Bremen, Bremen, Germany, e-mail: j.delhey@jacobs-university.de

Hyunji Kim Department of Psychology, University of Toronto Mississauga, Mississauga, ON, Canada, e-mail: hh.kim@utoronto.ca

Michael Koch Department of Neuropharmacology, Center for Cognitive Sciences COGNIUM, Brain Research Institute, University of Bremen, Bremen, Germany, e-mail: michael.koch@uni-bremen.de

Hanna Kokko Research School of Biology, ANU College of Medicine, Biology & Environment, Australian National University, Acton, Australia, e-mail: hanna.kokko@anu.edu.au

Morten L. Kringelbach Department of Psychiatry, University of Oxford, Oxford, UK, e-mail: morten.kringelbach@psych.ox.ac.uk

Christian Kroll Bremen International Graduate School of Social Sciences, Bremen, Germany, e-mail: c.kroll@jacobs-university.de

Alexander C. Pacek Department of Political Science, Texas A&M University, College Station, TX, USA, e-mail: e339ap@polisci.tamu.edu

Aloys Prinz Münster School of Business and Economics, Institute of Public Economics, University of Münster, Münster, Germany, e-mail: A.Prinz@wiwi.uni-muenster.de

Benjamin Radcliff Department of Political Science, University of Notre Dame, Notre Dame, IN, USA, e-mail: radcliff.1@nd.edu

Ulrich Schimmack Department of Psychology, University of Toronto Mississauga, Mississauga, ON, Canada; Valerie Tiberius Department of Philosophy, University of Minnesota, South Minneapolis, MN, USA, e-mail: uli.schimmack@utoronto.ca

Valerie Tiberius Department of Philosophy, University of Minnesota, South Minneapolis, MN, USA, e-mail: tiberius@umn.edu

Heinz Welsch Department of Economics, Carl von Ossietzky Universität, Oldenburg, Germany, e-mail: welsch@uni-oldenburg.de

Peter C. Whybrow Jane & Terry Semel Institute for Neuroscience and Human Behavior, University of California (UCLA), Los Angeles, CA, USA, e-mail: pwhybrow@mednet.ucla.edu

Song Yan School of Humanities and Social Sciences, Jacobs University Bremen, Bremen, Germany

Chapter 1
Happiness and Maximization: An Introduction

Hilke Brockmann and Jan Delhey

If you ask a chef, a physician, or a teacher the question: "Is more *always* better," they will probably answer: "No". Of course, it depends on the dish you are cooking, the illness you are curing, and the subject you are teaching. But these professionals know when additional ingredients spoil the dish, additional treatment harms the patient, and additional learning material frustrates the student. If you put the same question to an economist or a consumer, though, it is less clear what the answer will be.

Nowadays, consumers seem to have an insatiable desire for more and more goods and services. Economists typically equate more consumption with greater well-being. And growth has become a leading commercial and political imperative. As a consequence, Frank and Cook (1995) have described advanced societies as "winner-take-all-societies" in which a few superstars are paid handsomely, and others are seduced into mimicking their grand lifestyles (Frank and Cook 1995). In this environment, maximization of consumption has become the hallmark of success. More is universally taken to mean better, but here we ask whether more actually makes people happier.

To answer this question, we reassess the growth imperative with the help of subjective happiness measures. This is a risky venture since we enter trans-disciplinary terrain stretching from economics to biology and neuroscience. Nevertheless we believe it is a worthwhile one. For two reasons: A critical examination of the maximization principle permits a systematic exploration of the pleasures and pains of contemporary human life. At the same time, the growing empirical research on happiness and well-being in the social and natural sciences provides fresh insights into ways in which positive experiences may be created, nurtured, and made to contribute to the creation of happy lives. In this book, we test the critical potential of happiness findings to challenge the assumption that more is always better.

H. Brockmann (✉) · J. Delhey
School of Humanities and Social Sciences, Jacobs University Bremen, Bremen, Germany
e-mail: h.brockmann@jacobs-university.de

H. Brockmann and J. Delhey (eds.), *Human Happiness and the Pursuit of Maximization*, Happiness Studies Book Series, DOI: 10.1007/978-94-007-6609-9_1,

1.1 The Issue

Why should we be scrutinizing economic growth and maximization behavior at this particular point in time? From a historical perspective and on a global scale, human life has improved with rising affluence. Today, more people lead longer, safer, and more comfortable lives than ever before, despite the current economic crisis in Europe and America (Oeppen and Vaupel 2002). Globally, poverty and inequality have shrunk due to economic growth in China and India, particularly since the 1980s (Hung and Kucinskas 2011; Wade 2004).

However, while *average* material living standards have kept rising, signs of market failure have accumulated in affluent countries. Evidence suggests that an increasing number of people in Western countries are dropping out of what we might call the "comfort zone". In the OECD world, social inequality has increased sharply (Alderson and Nielson 2002). Nowadays, younger cohorts have to cope with a more versatile and competitive work environment than that of their parents. Labor markets, in particular, have transformed and undermined stable career patterns and secure economic prospects (Hacker 2006; Kurz et al. 2005). Most importantly, the young increasingly suffer from a lack of jobs. In Greece and Spain, for example, half of young adults under 25 were unemployed in July 2012 (Eurostat 2012).

Pundits highlight negative social and environmental externalities, costs or transaction spillover, of contemporary capitalism. In their view, the drive to maximize profit not only boosts incomes and wealth inequalities, but also boosts crime rates, family breakups, environmental pollution, and negative health trends (Porritt 2007; Stiglitz 2012). The World Wildlife Fund (WWF) claims that the ecological footprint of an average high-income country is five times that of a low-income country.[1] They estimate that Mankind as a whole would currently need 1.5 Earths to live ecologically sustainably, and 2 Earths by 2030 (WWF 2012).

The observed increase in waistlines globally also indicates that people are living beyond their means. The rich United States is home to the largest proportion of obese people in the world (WHO 2011; Wisman and Capehart 2010). A super-sized economy apparently produces super-sized citizens. In addition, there is mounting evidence that symptoms of anxiety and depression are on the increase, particularly among the young. This is the flipside of capitalism's meritocratic culture (Collishaw et al. 2004; Twenge et al. 2010). The current global financial crisis has afflicted people with depressive symptoms most severely (Catalano et al. 2011; Lee et al. 2010) and increased the consumption of antidepressants (Olfson and Marcus 2009). Psychological problems are now the number one reason for sick-leave in many Western countries.

At the same time, a growing number of people in affluent societies are questioning the single-minded pursuit of material maximization and conspicuous consumption. Inglehart (1997) puts this "culture shift" into the broader context of

[1] You can calculate your own ecological footprint at: http://www.myfootprint.org/.

post-modernization. With emerging post material values and alternative life styles, people are aiming at a new quality of life. They are creating new movements for slow food, fair trade, self-sustaining communities and green politics. As Delhey (2010) has shown, the ingredients for a satisfying life—the "happiness recipe" as he calls it—is less materialistic in affluent societies than in poorer countries.

All these developments create the suspicion that rich nations may have reached a historical turning point where the capacity for economic growth to produce higher well-being has largely or completely been exhausted (Ayres 1998; Wilkinson and Pickett 2009). The Genuine Progress Index, based on the idea that GDP has to be offset by an allowance for ecological pollution and social costs, shows no increase in the US for the past 30 years, despite an impressive growth in standard GDP. The authors conclude that bigger isn't necessarily better (Cobb et al. 1999). But how robust is this finding. Do other measures show the same degree of stagnation, or perhaps even a decline in quality of life in affluent societies? Are people happier with less?

The main unit of analysis throughout the book is therefore the individual. Her and his well-being and happiness, empirically verified, should provide us with more detailed insights into the benefits and costs of wealth-maximizing regimes. However, in three chapters we extend this. We descend to the level of individual brain structures and extend to the level of entire non-human populations. This allows us to take account of new findings from the natural sciences. It is expected that an understanding of the biological mechanisms underlying maximizing behavior and the experience of happiness will provide us with a firmer basis for our beliefs.

First, the desire for more of everything appears to be ingrained in human nature. We all want more appreciation, more convenience, more excitement, and more choice. The benefits of this desire have long been emphasized by many philosophers and social scientists. The founding fathers of Utilitarianism raised the maximization of utility to the status of a fundamental ethical principle. Yet, the utilitarian credo of "the greatest good for the greatest number" (Bentham 1780/2007, p. 1) did not become the guiding principle of capitalist societies in its original sense, as the maximization of happiness. It became vulgarized, and came to mean "the greatest income for the greatest number"—that is the maximization of wealth. Generations of economists have tended to equate material progress with social progress. Political scientists and sociologists also often assumed that humans are motivated by the desire for more power and more social recognition (Coleman 1990; Mills 1956). Other sciences have followed a similar line of reasoning. Biologists have emphasized a presumed evolutionary imperative to maximize the spread of our genes and calorie intake (Krebs and Davies 1993; Parker and Smith 1990). Neuroscientists and decision scientists have tended to evaluate experimental outcomes against maximizing strategies (Dayan and Abbott 2005; Grant and van Zandt 2009; Savage 1954)

Yet, from the beginning of what we might term modernity, the maxim that more is always better has also faced opposition. Malthus, for example, predicted that starvation would be a consequence of unchecked growth (Malthus 1789/2008).

Jevons saw that technological progress would result in the exhaustion of coal deposits (Jevons 1866/2009). And in 1972, the Club of Rome published its famous ecological study entitled "The limits of growth" (Meadows et al. 1972).

Although many of these critical growth predictions, in particular the older ones, proved to be unsound, the issue of ecological sustainability is still with us and remains unsolved, as the recent debate on planetary boundaries demonstrates (Rockström et al. 2009; Victor 2010). The issue constitutes a great challenge to the ideas of limitless growth and material maximization.

The new science of *happiness* (Layard 2005) complements these objective critiques of maximizing strategies and ecological exploitation with a subjective and individual-centered perspective. Biologists, economists, neuroscientists, political scientists, psychiatrists, psychologists, and sociologists have looked more deeply into the motives driving human behavior. Together, they have produced insights into what matters for people in their lives—insights that Frey (2008) has termed "revolutionary".

The crucial question in our context is whether material maximization makes people happier in affluent societies. Rather than letting experts decide on this matter, happiness researchers use individuals' self-reported or self-indicated appraisal as indicators. They follow the simple idea that any good, activity, or event is desirable which increases the subjectively experienced quality of life. This implies that more is better. Nevertheless, the pursuit of happiness does not exclude the possibility of decreasing marginal utility, or an inverse utility function, so at a certain threshold, more could still detract from well-being.

Indeed, a few happiness researchers claim that life in affluent societies is actually detrimental to feelings of well-being. One argument is that affluence encourages people to focus too much on money and success, so that their more fundamental needs of belonging and leading a meaningful life are not met (Bannas 1989). Another is that material aspirations have grown to such an extent that we are unable to value properly what we have, so that dissatisfaction prevails (Wachtel 1989). The implication of this radical position is that we were better off, in the sense of being happier, when we had *less*.

There is empirical support for the proposition that getting richer does not make us any happier. It is called the Easterlin paradox. Easterlin (1974, 1995) showed that richer countries are on average happier than poorer ones but that, beyond a certain point, further economic growth within a country does not improve the people's sense of well-being because people adapt quickly to improvements in living standards. This paradox was first observed in the US and found to hold also for Japan; but it was found to hold less for Europe. Recent studies found even small but significant increases in life satisfaction in a number of European countries (Hagerty and Veenhoven 2003; Veenhoven and Hagerty 2006). For example, in Ireland happiness has increased largely at times of the economic boom (Madden 2011). Some analysts contest the Easterlin paradox even for the prime example of the US (Fischer 2008). So, the country-level studies in this book put the Easterlin paradox to the test.

Easterlin himself relates his findings to individual-level processes such as habituation and adaptation. Much empirical research associated with it aims at understanding and decomposing individual happiness into its elements. Issues addressed include people's expectations and aspirations, their decision processes, actual choices, and their reactions towards major life events and changing living conditions (Brockmann 2010; Frijters et al. 2011; Gilbert and Ebert 2002; Kahneman et al. 1999; Schwartz 2005; Suls et al. 2002; Zimmermann 2007). So far, it is a well-established fact that income has a decreasing marginal utility, and that we adapt at least partly to income changes (Clark 1999; Frijters et al. 2004). But it is less clear if this applies to non-monetary sources of happiness, too.

Finally, happiness as a measure of well-being adds a completely new dimension to the maximization debate: Is it *happiness* we should aim to maximize or something that makes us happy? Under the tenets of classical utilitarianism it was self-evident that societal happiness should the ultimate goal, and that improving it was the foremost duty of the government. But what when current empirical research shows that we cannot achieve happiness directly? Moreover, both claims are contested among philosophers. Some argue that well-being should be more than mere happiness (Tiberius 2013 in Chap. 5), and many regard happiness as a private matter in which the state should not interfere. Obviously, these are normative questions that cannot be resolved by empirical data alone. However, it is also clear that any proposed solution should be informed by such data.

1.2 The Book and its Chapters

Our book focuses on contemporary affluent societies. Economic growth here is still the policy priority for governments, and higher income a major aim for individuals. The maxim that "more is better" seems still generally accepted. But critics identify limits of growth in objective and social externalities. We add subjective costs to the equation. Returning to the initial utilitarian ideal, all authors in this book, despite their diverse background disciplines, apply happiness as a yardstick, defined and measured as the subjective evaluation of a person's overall quality of life (Veenhoven 1984). Understood in this way, happiness is a universal, cognitive and emotional self-evaluation process of environmental stimuli (Veenhoven 2010) it motivates behavior and may help to decide when more is better, and when enough is enough. Happiness measures employed in the book capture this subjective self-evaluation through questionnaires, evolutionary success, brain activity, or even molecular reactions.

In the first chapter, the US takes center stage. *Peter Whybrow* brings a medical perspective to the case of a money-driven and success-oriented country. As a psychiatrist, he is concerned about the increasing amount of mental ill-health amongst Americans. In contrast to most medical professionals he reflects on the systematic impact that a competitive and materialistic social environment has on the psyche of many people. He identifies a substantial gap that exists between human biological predispositions on the one hand, and the requirements of maximizing markets on the other. The author has dubbed this the "American experiment", but points out that it is not an "experiment" that has been conducted in the lab, but in real life. He asks whether other affluent societies will follow the same track. There is a risk that globalization will spread the American-style imbalance between human needs and capitalist markets to other corners of the globe. Yet, according to Whybrow, the path towards ill-being may be reversed if Americans were to live a more balanced life in which they focused less on materialism, and more on social relations and productive social engagement.

Matthias Binswanger adds another critical diagnosis of affluent societies in Chap. 3—artificial competition. As an economist, he is skeptical about the expansion of quasi-markets into every sphere of public life. He illustrates his argument with many examples, and shows that in areas such as education, health and science, there is no market in the classical sense; hence the artificial competition introduced in these areas cannot work properly. With a hint of irony, Binswanger describes the sophisticated, but ultimately hopeless, attempts of controllers to measure "quality" with quantitative indicators. Yet just as in the fairy tale of the sorcerer's apprentice, the numbers develop a life of their own and create perverse incentives for social actors so that competition becomes an end in itself. The consequence is a decrease rather than an increase in public welfare, and a decrease in the individual happiness of both producers and consumers.

The subsequent two chapters discuss the idea of maximization from a philosophical perspective. In Chap. 4, *Kurt Bayertz* is concerned with the requirement to maximize outcome in a simplistic and prescriptive way. He starts by distinguishing between maximization as a moral principle (where the imperative is to promote overall happiness in a society) and maximization as a prudential principle (where the imperative is to promote the agent's own happiness), and focuses on the latter. Drawing on empirical findings and a thought experiment, Bayertz contends that maximization is not conducive to the agent's happiness. But Empirical studies suggest that maximizers face greater decision-related stress and tend to be less satisfied with their decisions. This arises because of the insecurity engendered about one's own preferences. The potential endlessness of decision processes, and the impact of external constraints, prevent agents from developing a clear-cut and satisfying maximization strategy.

In Chap. 5, *Valerie Tiberius* addresses the crucial question as to whether we should aim to maximize happiness or something else. Departing from classical Utilitarianism as espoused by Jeremy Bentham and John Stuart Mill, she makes a conceptual distinction between feeling happy and leading a good life. Tiberius argues that the "all-encompassing good" is the only candidate for maximization, and that, although happiness is part of this, there is something else. This pluralism has consequences for the idea of maximizing the "all-encompassing good". Whereas some of its ingredients are quantifiable and hence can be maximized (income, for example), others are not quantifiable and hence cannot be easily maximized (friendship, for example). This pluralism also means that we are often faced with having to make trade-offs between things we value, so that maximizing one good will always be constrained by other goods that contribute to happiness. Within these constraints, however, trying to produce more of some things we value still makes good sense. So, gaining wealth can still be a good strategy to achieve what one values. But it is neither always the right strategy, nor is it the only way to build a good life.

The claim that a loose coupling exists between maximization and happiness is also supported by empirical studies. Chapter 6 by *Hannah Kokko* provides us with examples from evolutionary biology that portray a seemingly "unwise" mother nature. Her key message is that there are natural non-competitive environments that do not require maximization strategies to secure the survival of species. Only when resources are scarce and fixed do species have to play a zero-sum game in which nature favors competitive strategies and guarantees only the survival of the fittest. Also, there are examples when a reproductive success turns out to be a Pyrrhic victory, because initial growth may ruin an entire population in the long run. Kokko draws her examples from animal kingdom (e.g., kestrels, Amazon mollies). But the game theoretic logic has been modelled on human behavior, so it is tempting to apply her observations to human populations.

Whereas evolutionary biologists work on a large scale, surveying huge structures and very long-term developments, neuroscientists work on a small-scale, typically surveying small brain structures and neural circuits. During the last few decades neuroscientists have been able to shed light on human maximization behavior and their experience of pleasure by analyzing brain activity. In Chap. 7, *Kent Berridge* and *Morten Kringelbach* examine hedonic brain mechanisms and show that human neural networks associated with "higher" pleasures overlap with those associated with basic sensory pleasures. So they posit the existence of an overall hedonic brain circuitry which comprehends both very concrete and more abstract experiences involved with the sensation of well-being. Brain scans illustrate how hedonic experiences are permanently processed in phases of wanting, liking and learning. The central location of this hedonic circuitry is explained by the key role pleasures play in decision making processes essential for survival. Further, perceiving pleasure as a system of wanting, liking and learning reframes the problem of maximizing happiness to a problem of balancing different aspects of hedonism.

Could this balance be improved? In Chap. 8, *Michael Koch* explains how progress in psychopharmacology over recent years has greatly increased the possibilities of treating various forms of mental dysfunction and of enhancing cognitive performance and happiness in general. The chapter reviews the neurobiological basis of drugs that are currently used to treat neuropsychiatric disorders such as anxiety, depression, schizophrenia, Alzheimer's disease, attention deficit hyperactivity disorder (ADHD) and addiction, and outlines how these drugs might be used in healthy persons to improve mental functioning or emotional states. The chapter closes by addressing some social, legal and ethical concerns associated with "brain doping".

Ulrich Schimmack and *Hyunji Kim* discuss in Chap. 9 the issue of psychological adaptation, which is a headache for neuro-enhancement and is more generally an obstacle to improving levels of happiness. According to conventional psychological wisdom, people adapt quickly to changing life circumstances, so that getting happier is almost impossible. Schimmack and Kim challenge this idea of a "hedonic treadmill" by pointing out both conceptual flaws in the set-point theory, and flaws in how results from studies conducted within the set-point paradigm are interpreted. The authors suggest that life events and changing living conditions *do* have a lasting impact on how people feel about their lives, much more so than conventional psychological wisdom holds. Does that mean that positive psychology is right in claiming that happiness levels can be easily lifted, either by means of more prudent individual strategies or public policies conducive to happiness? The authors argue that, for the affluent Western societies, the answer is "No". This is for two simple reasons. People are already quite happy in these places, and subjective well-being is known to have a strong genetic basis, which is typically normally distributed and cannot be "improved". Schimmack and Kim conclude that there is likely to be room for a small improvement in average happiness.

From the *individual* treadmill of hedonic adaptation, Chap. 10 by *Hilke Brockmann* and *Song Yan* shifts attention to the potential *social* treadmill of social comparison and positional goods. Drawing on a wide range of literature and data on consumption and subjective well-being, they show that in affluent societies consumers tend to invest too much in "keeping up with the Joneses". The resulting over- or malconsumption does not lead to greater average happiness. Increasing income inequality in most OECD countries, and in emerging economies during the last 30 years, has fuelled myopic shopping behavior which seduces consumers into overspending on positional goods. In line with recent happiness findings, Brockmann and Yan point out that consumers could better gain well-being by buying more experiences instead of goods, by sharing things, by delaying gratifications, and by re-evaluating the consumption of fellow citizens as a positive sign towards one's own future options. At an institutional level, the authors suggest that curbing income inequality, and taxing conspicuous consumption more heavily, could help nudge consumers away from excessive and counter-productive competition for status.

Chapter 11 by *Heinz Welsch* adds new insights from happiness research to the issue of environmentally sustainable consumption. Welsch demonstrates that the economic analysis of happiness has identified previously neglected routes by which current consumption behavior negatively affects environmental quality, *over and above* externalities such as pollution. He further argues that current consumer choices are also suboptimal from the perspective of inducing individual happiness; consumers would be happier if they were to buy more environmental-friendly goods. This is because, in their decision-making, humans tend to be biased towards extrinsically motivated options and away from intrinsically motivated ones. Buying "green" belongs to this latter category. Nevertheless, Welsch's encouraging message is that progress towards ecologically sustainable lifestyles would kill two birds with one stone; it would help the environment as well as engendering human happiness. Contrary to conventional wisdom, there is no inherent trade-off between the two.

In Chap. 12, *Robert Davidson*, *Alexander Pacek* and *Benjamin Radcliff* draw our attention away from individual consumer choice and towards public choice and the welfare state. Their research asks to what extent state intervention in market outcomes is conducive to subjective well-being. By using life satisfaction as the dependent variable, they provide robust evidence that de-commodification—a summary measure of how much the welfare state shields people from market forces—and labor market regulations such as on-the-job safety and lay-off protection, contribute to happiness in an OECD-wide comparison. They further provide qualified evidence that government size has a similar positive effect on average happiness. Together these results clearly contradict the idea that unfettered markets, the famous "invisible hand", are the most efficient means of creating human well-being. This does not imply that state coordination trumps market coordination, since all OECD countries *are* market economies; rather, the key message is that subjective well-being is maximized when fully-fledged market economies are accompanied by a strong and effective welfare state, so that market deficits are compensated for, and citizens can benefit from the best of both worlds—the world of economic freedom, and the world of economic security.

In Chap. 13, *Aloys Prinz* investigates whether the state should care for the happiness of its people, and if it should, how it should best go about this. Reviewing arguments from proponents and opponents of the concept of the "caring state", he argues that the state should not feel responsible for the happiness of its citizens *directly*, since happiness is an individual matter; rather, the state should feel responsible for creating the *background conditions* that are most likely to make people happy. The author asks whether there is any form of welfare state that is best suited to achieving this. By considering both individual autonomy and security as two basic pillars of human happiness, Prinz claims that the ideal state would be one which balances autonomy and security and that an "avuncular state" can achieve the best balance. Happiness would increase when citizens were nudged by the state to make better choices, but were allowed to be more involved in the political decision making of the state.

Politicians who care about national well-being and happiness need precise data and statistics on what makes citizens' life worthwhile. Starting from Kennedy's famous dictum about the inappropriateness of the GDP in this regard, the final chapter, Chap. 14 by *Jan Delhey* and *Christian Kroll* discusses both the GDP and new, alternative welfare measures and puts them to the test. They ask which index of national well-being is best suited to explain international differences in happiness in the OECD world. The results are both sobering and encouraging for the social indicators movement. The sobering result is that the GDP is a much better indicator of a nation's happiness than many think, and most new welfare measures do not outperform the GDP in the happiness test. The encouraging result is that there is one index—the OECD's newly launched Better Life Index—that does do a better job in the happiness test. Delhey and Kroll conclude by explaining why they think the Better Life Index performs well, and suggest how this might contribute to the design of alternative welfare measures.

1.3 In Conclusion: More Must be Better Balanced

Throughout the book, philosophers, social scientists, and natural scientists provide us with ideas, data, and examples showing when, where, and why, more is not always better. The new science of happiness provides the basis for integrating findings from evolutionary biology, neuroscience, psychiatry, philosophy, political science, sociology, and economics. This does not create a single coherent theory, but the results do provide compelling evidence that maximization and happiness are only *loosely coupled* in determining human and non-human behavior.

At a country level, we do find that people in richer nations are on average happier and more satisfied than those in poorer ones. But, in line with the Easterlin paradox, beyond a certain point, additional growth does not make us much happier, if at all. More is good but not necessarily better, because we either adapt, compare ourselves too much with others, or have to cope with the social costs of economic growth which can cancel out the happiness gained from more income. In the book you will find accounts of situations in which each of the three influences (adaptation, comparison, cost) has an impact.

Clearly, a happy life is about more than economic success. The happiest nations are those which are not only economically well-off, but safe, non-corrupt, and possess a high stock of social capital. The power of this pluralism of social conditions to increase happiness needs to be acknowledged by policy makers, and it should also be recognized when developing new indicators of well-being that are needed to guide political and other experts in their decision making. The fixation on GDP alone can no longer be justified. Likewise, at the level of the individual, happy men and women are typically those doing well in all three areas of life, that is in having, loving, and being (Allardt 1993).

The pluralism in the sources of happiness takes us back to the idea of *balancing* various social influences. This idea is shared by many authors of the book and it is backed by biological and neuroscience findings. The hedonic circuitry in the brain illustrates why we cannot maximize happiness or liking directly, but have to do so indirectly via the stages of wanting and learning. These findings answer the philosophical question, whether we should aim at happiness or something that makes us happy. Future research may also prove that these physical boundaries define when enough consumption is enough.

Today, happiness research teaches us the benefits to be had from placidity and resilience in a world of global competition, high-flying aspirations, and accelerated production cycles. One does not need to be the richest, most-liked, and most-dedicated person in order to be happy. In most situations, and for most people, it is satisfying to have a good income, to be liked, and to have some purpose in life. Ordinary life is not a sprint but a marathon which demands great staying power, smart resource management, and the adaptation of expectations to an ever changing environment.

The concept of balancing also captures better the different early warning signs of a maximizing economy. Research identifies new imbalances in what has become a 'winner-takes-all' society, many happy people alongside with increasing numbers of unhappy people; the young unemployed, the sick middle-aged, and poor single mothers, who all face harsh competition, insecurity, and social exclusion. Furthermore, empirical happiness research provides tools to disaggregate averages and to disentangle the complex dynamics of subjective well-being over time (Brockmann and Delhey 2010). A more fine-grain analysis is better able to accommodate new outliers and critical episodes to meet the "challenge of affluence" (Offer 2006).

At the same time happiness findings also help indicate where there is room for improvement and how that need might be met. Authors of this book suggest institutional, structural, and cultural reforms which curtail competition and social inequality in order to nudge people to make smarter, long-term, and less selfish decisions, to overcome unhealthy upward comparisons and to promote a more solidary culture. The authors also agree that individuals would benefit from living a more balanced life in which we try not too hard to maximize only one aspect in life but strive for a healthier work-life balance.

In a nutshell, happiness research can guide us in our attempt to collectively organize our societies more wisely, and to make individual choices more prudently. In theory, there is less room to increase happiness at the high end of the scale than at the low end. But, as individuals, we know that our life has both highs and lows. And even though we often stress throughout the book how difficult it is to maximize subjective well-being, we believe that progress in Jeremy Bentham's original sense is both necessary and possible. We hope that this edited volume provides both impetus and substance to the debate and thus contributes to reaching the real utilitarian goal of providing for a society the greatest happiness for its greatest number.

Acknowledgments We are indebted to many people and institutions for support and advice along the way.

Most chapters in this book originated from a workshop on "Happiness and Maximization", which the editors of this book organized in association with Kurt Bayertz. Uwe Opolka brought the three of us together and provided us with the material and mental resources to kick off the cooperative endeavour. His sudden death made us very saddened. We would like this book to serve as a memorial to an inspiring, very friendly and supportive colleague. The workshop was finally held in January 2011 at the Center for Advanced Studies/Hanse Wissenschaftskolleg (HWK) in Delmenhorst, Germany. We are very grateful to the HWK for hosting our workshop and for supporting it in a host of different ways. Generous financial support from the Fritz Thyssen Foundation in Cologne and the Bremen International Graduate School of Social Sciences (BIGSSS), a common enterprise of University Bremen and Jacobs University Bremen, is also acknowledged with thanks.

We further would like to thank Antonella Delle Fave, chief-editor of the *Journal of Happiness Studies*, for being so supportive of our project and for accepting its product in the newly launched *Journal of Happiness Studies* book series. We also received a lot of encouragement and support from Springer. In particular we want to thank Ester Ottens for guiding us through the different stages of the project. We are obliged to Jacobs University and its Research Center "Cognition, Values and Well-being" (CoWell) for financial support in finishing the book. A special thanks goes to Katja Uglanova for helping us with many technical and organizational matters in completing the manuscript. Finally, we would like to thank all authors for their valuable contributions and their positive attitude towards this book.

Bremen, October 2012

Hilke Brockmann and Jan Delhey

References

Alderson AS, Nielson F (2002) Globalization and the great U-turn: income inequality trends in 16 OECD countries. Am J Sociol 107:1244–1299

Allardt E (1993) Having, loving, being: an alternative to the Swedish model of welfare research. In: Nussbaum M, Sen A (eds) The quality of life. Clarendon Press, Oxford, pp 88–94

Ayres R (1998) Turning point: an end of the growth paradigm. Earthscan, London

Bannas S (1989) Reich—dafür unglücklich? Transfer-Verlag, Regensburg

Bentham J (1780[2007]) An introduction into the principles of morals and legislation. Clarendon Press, Oxford (unabridged Dover edition)

Brockmann H (2010) Why are middle-aged people so depressed? Evidence from West-Germany. Soc Indic Res 97:23–42

Brockmann H, Delhey J (2010) The dynamics of happiness. Soc Indic Res 97:1–5

Catalano R, Goldman-Mellor S, Saxton K, Margerison-Zilko C, Subbaraman M, LeWinn K, Anderson E (2011) The health effects of economic decline. Annual review of public health, 32:431–450

Clark AE (1999) Are wages habit-forming? Evidence from micro data. J Econ Behav Organ 39:179–200

Cobb C, Goodman GS, Wackernagel M (1999) Why bigger isn't better: the genuine progress indicator: 1999 update. Redefining Progress, San Francisco

Coleman JS (1990) Foundations of Social Theory. Cambridge, MA: Harvard University Press

Collishaw S, Maughan B, Goodman R, Pickles A (2004) Time trends in adolescent mental health. J Child Psychol Psychiatry 45:1330–1362

Dayan P, Abbott LF (2005) Theoretical Neuroscience. Computational and Mathematical Modeling of Neural Systems. Cambridge, MA: MIT Press

Delhey J (2010) From materialist to post-materialist dappiness? National affluence and determinants of life satisfaction in cross-national perspective. Soc Indic Res 97:65–84

Easterlin RA (1974) Does economic growth improve the human lot? Some empirical evidence. In P.A. David & M. W. Reder (Eds.), Nations and households in economic growth: Essays in honor of Moses Abramaowitz. New York: Academic Press, pp 89–125

Easterlin, R. A. (1995). Will raising the income of all increase the happiness of all? Journal of Economic Behavior and Organization 27:35–48

Eurostat E (2012) Euroindicators. News release. Available at https://epps.eurostat.ec.europa.eu/portal/euroindicators/news_releases. Accessed on 12 July 2012

Fischer C (2008) What wealth-happiness paradox? A short note on the American case. J Happiness Stud 9:219–226

Frank RH, Cook PJ (1995) The winner-take-all society. The Free Press, New York

Frey BS (2008) Happiness: a revolution in economics. MIT Press, Cambridge

Frijters P, Haisken-DeNew JP, Shields MA (2004) Money does matter! Evidence from increasing real income and life satisfaction in East Germany following reunification. Am Econ Rev 94:730–740

Frijters P, Johnston DW, Shields MA (2011) Life satisfaction dynamics with quarterly life event data. Scand J Econ 113:190–211

Gilbert DT, Ebert JEJ (2002) Decisions and revisions: the affective forecasting of changeable outcomes. J Pers Soc Psychol 82:503–514

Grant S, van Zandt T (2009) Expected utility theory. In: Anand P, Pattanaik PK, Puppe C (eds) The handbook of rational and social choice. Oxford University Press, Oxford, pp 21–68
Hacker JS (2006) The great risk shift: the new economic insecurity and the decline of the American dream. Oxford University Press, Oxford
Hagerty MR, Veenhoven R (2003) Wealth and happiness revisited: growing national income does go with greater happiness. Soc Indic Res 64:1–27
Hung HF, Kucinskas J (2011) Globalization and global inequality: assessing the impact of the rise of China and India, 1980–2005. Am J Sociol 116:1478–1513
Inglehart R (1997) Modernization and postmodernization: cultural, economic, and political change in 43 societies. Princeton University Press, Princeton
Jevons WS (1866 [2009]) The coal question: an inquiry concerning the progress of the nation, and the probably exhaustion of our coal mines. Kessinger Publishing, Whitefish
Kahneman D, Diener E, Schwartz N (1999) Well-being: the foundations of hedonic psychology. Russell Sage, New York
Krebs JR, Davies NB (1993) An introduction to behavioural ecology, 3rd edn. Blackwell, Oxford
Kurz K, Blossfeld H-P, Klijzing E, Mills M (2005) Globalization, uncertainty and youth in society. Routledge, London
Layard R (2005) Happiness: lessons from a new science. Penguin, New York
Lee S, Guo W-J, Tsang A, Mak ADP, Wu J, Ng KL, Kwok K (2010) Evidence for the 2008 economic crisis exacerbating depression in Hong Kong. J Affect Disord 126:125–133
Madden D (2011) The impact of an economic boom on the level and distribution of subjective well-being: Ireland 1994–2011. J Happiness Stud 12:667–679
Malthus TR (1789[2008]) An essay of the principle of population. Oxford University Press, Oxford
Meadows DH, Meadows DL, Randers J, Behrens WW (1972) The limits to growth. Universe Books, New York
Mills CW (1956) The Power Elite. Oxford: Oxford University Press
Oeppen J, Vaupel JW (2002) Broken limits to life expectancy. Science 296:1029–1031
Offer A (2006) The challenge of affluence: self-control and well-being in the United States and Britain since 1950. Oxford University Press, Oxford
Olfson M, Marcus SC (2009) National patterns in antidepressant medication treatment. Arch Gen Psychiatry 66:848–856
Parker GA, Smith JM (1990) Optimality theory in evolutionaly biology. Nature 348:27–33
Porritt J (2007) Capitalism as if the world matters. Earthscan, London
Rockström J, Steffen W, Noone K, Persson Å, Chapin FS III, Lambin EF et al (2009) A safe operating space for humanity. Nature 461:472–475
Savage LJ (1954) The foundations of statistics. Wiley, New York
Schwartz B (2005) The paradox of choice. Why more is less. HarperCollins, New York
Stiglitz JE (2012) The price of inequality. Norton, New York
Suls J, Martin R, Wheeler L (2002) Social comparison: Why, with whom, and with what effect? Curr Dir Psychol Sci 11:159–163
Twenge JM, Gentile B, DeWall CN, Ma D, Lacefield K, Schurtz DR (2010) Birth cohort increases in psychopathology among young Americans, 1938–2007: a cross-temporal meta-analysis of the MMPI. Clin Psychol Rev 30:145–154
Veenhoven R (1984) Conditions of happiness. D. Reidel Publishing Company, Doordrecht
Veenhoven R (2010) How universal is happiness. In: Diener E, Kahnemann D, Helliwell JF (eds) International differences in well-being. Oxford University Press, New York, pp 328–350
Veenhoven R, Hagerty MR (2006) Rising happiness in nations 1946–2004: a reply to Easterlin. Soc Indic Res 79:421–436
Victor PA (2010) Managing without growth: slower by design, not disaster. Edward Elgar, Cheltenham
Wachtel PL (1989) The poverty of affluence: a psychological portrait of the American way of life. New Society Publishers, Philadelphia
Wade RH (2004) Is globalization reducing poverty and inequality? World Dev 32:567–589

WHO (2011) Global Infobase. Available at https://apps.who.int/infobase/Index.aspx. Accessed on 12 July 2012

WWF (2012) WWF Living Planet Report. Available at http://awsassets.panda.org/downloads/1_lpr_2012_online_full_size_single_pages_final_120516.pdf. Accessed on 12 July 2012

Wilkinson R, Pickett K (2009) The spirit level: why more equal societies always do better. Penguin Books, London

Wisman JD, Capehart KW (2010) Creative destruction, economic insecurity, stress, and epidemic obesity. Am J Econ Sociol 69:936–982

Zimmermann AC (2007) Adaptation, assets, and aspirations. Three essays on the economics of subjective well-being. Doctoral dissertation. University of Southern California, Los Angeles

Chapter 2
Is More Always Better? The American Experiment

Peter C. Whybrow

They (Americans) find prosperity almost everywhere, but not happiness. For them desire for well-being has become a restless, burning passion which increases with satisfaction.
(Alexis de Tocqueville 1835, Democracy in America, vol. I)

2.1 In Pursuit of Happiness

In the United States happiness is a time honored and hallowed pursuit, one enshrined in *The Declaration of Independence* along with liberty and life itself. To achieve happiness is the cultural equivalent to King Arthur discovering the Holy Grail: it is a sacred quest.

Americans are an optimistic, can-do people possessed of burning ambition. This extraordinary drive is part of the migrant temperament, as I have explored in earlier writings (Whybrow 2006). Drawing upon this reservoir of restless vigor the US has built a unique culture, one that is held together by a commitment to individual freedom and progressive material betterment. In this quest for more it is also a culture that promotes an evangelical individualism. Thus, in the nineteenth century, the popular Horatio Alger's rags to riches stories of courage and hard work grew out of the simple notion that only in America can the future be grasped and made one's own. This remains the founding mythology: that the pain of the present is to be endured for it is the future that holds the key to happiness, moving up the economic ladder to secure a better life for oneself, for the family and for the children. It is a vision of the future made manifest through social mobility and the maximizing of material gain: it is the American Dream.

Today, for many Americans the Dream isn't what it used to be. There's a pervasive sense of unease. The citizens of the world's richest nation are beginning to feel that there should be more to life than two jobs and a flat paycheck. It's not that Americans have lost their drive. Even during the dark days of the financial crisis in 2009 surveys confirmed that over two thirds of citizens still believed that skill and hard work are the main ingredients for success and life-time happiness

P. C. Whybrow (✉)
Jane and Terry Semel Institute for Neuroscience and Human Behavior,
University of California (UCLA), Los Angeles, CA, USA
e-mail: pwhybrow@mednet.ucla.edu

H. Brockmann and J. Delhey (eds.), *Human Happiness and the Pursuit of Maximization*, Happiness Studies Book Series, DOI: 10.1007/978-94-007-6609-9_2,

(Upper bound 2010, p. 37–38). But many Americans report themselves increasingly anxious (Kessler et al. 2005). The country is changing and despite its dominance in the world they see dark times ahead. As a nation the US is carrying massive personal and government debt just at a time when the gusher of wealth seems to be drying up, along with the oil. In the first decade of the new century the average wage for the American worker has been stagnant with a widening gap between rich and poor (CBO 2011). Higher education is ever more expensive and the defining core of the Dream—America's much vaunted social mobility—is now largely a thing of the past. In a poll conducted for *The Economist* a third of the respondents believed that they had less opportunity to improve their standard of living than did their parents a generation earlier (Upper bound 2010, p. 37–38). And that perception is real. In 2004 men in their 30s earned 12 % less in real terms than did their fathers at a similar age, according to a Pew Foundation study on economic mobility (Meritocracy 2005, p. 22–24). Upward social progression is now greater within the European Union than it is within the fifty United States. America's self-image as the dreamland of betterment through maximization is under strain. So what is going on? To answer that question we must first look back to the philosophical roots of America's striving.

2.2 The Great Experiment

At its founding in the eighteenth century the United States of America was the Great Experiment in Enlightenment thinking—a democracy to be validated by the pursuit of individual freedom, initiative and hard work rather than by the exercise of arbitrary authority or religion. Thus Garry Wills, the distinguished American historian, has suggested in his book *Inventing America* (Wills 1978) that the construction of *The Declaration of Independence* reflects both the prevailing moral philosophy of the time and also the contemporary scientific preoccupations with Newtonian theory. It fell to Thomas Jefferson, who was well versed in the writings of John Locke, David Hume, and Adam Smith, among others, to speak eloquently for what the American colonists thought they were or could be.

Theirs was a grand vision. The touchstone for it all was the Enlightenment principle that through the exercise of human reason and the acquisition of knowledge through objective observation—a philosophy that is at the heart of science, law and the freedom of a parliamentary democracy—human potential could be maximized. Thus the *Declaration* was constructed as both a political and a moral document and Jefferson in his writing tried to capture this sentiment. In substituting the word happiness for property Jefferson was not confused. Rather in keeping with Adam Smith's vision he saw the protection of property as a central freedom in sustaining democratic ideals. The right to the exchange of property was grounded in Smith's free market principles and happiness flowed from the successful pursuit of such freedoms.

That the life in America was challenging, demanding physical stamina and mental ingenuity, complimented the philosophy of striving for continuous material betterment. Such cultural sentiment is to be found in Benjamin Franklin's autobiography (Lemay and Zall 1986)—and in his thirteen virtues to be pursued in the development of inner character—reflecting the drive for self-improvement that is still evident in the migrant mind of the American. Thus the Founding Fathers saw their nascent project as an experiment in how to live. It was in the minds of the leaders of the American Revolution that in exercising their freedom from Britain the colonies would not only validate these principles but also successfully implement them within the ideology of the free market to become a shining example to the rest of the world.

2.3 Adam Smith and Self-Regulating Markets

True to its founding, the United States of America remains the quintessential free-market society. But what does that mean? In human experience markets are ubiquitous, emerging wherever people congregate—along the riverbank, in the courtyard, and on the village street—and it has been ever so. But it was Adam Smith, the Scottish moral philosopher and the patron saint of capitalism, who in the nascent days of the American Republic championed the social value of harnessing the instinctual drives of curiosity and self-interest within the framework of the marketplace to create a self-regulating economic order. Adam Smith gave heft to our natural propensity for barter for as he described them, markets are the most "simple and obvious system of natural liberty."

Adam Smith was not a doctrinaire free trader, as he is frequently caricatured, but a careful student of human behavior who thought deeply about social issues. While self-interest drove the market Smith believed that its stability was grounded in the human propensity for compassionate collaboration with others and the need to be loved. This Smith called social sentiment: a socially acquired mental mechanism that goes beyond the ability to communicate one's own feelings to an understanding of how others feel, a capacity that today we call empathy. Within this dynamic social framework self-interest, appropriately shaped through the give-and-take of the market, made possible a society where the products of individual labor are fairly traded, placing a decent life within the reach of all. Thus in Smith's construct—as set forth in *The Wealth of Nations* in 1776—it was self-interest together with the instinctual drives of curiosity and ambition that fueled the engine of the marketplace while social feedback functioned as the brake to create a dynamic self-regulating social system (Smith 1776) (Fig. 2.1).

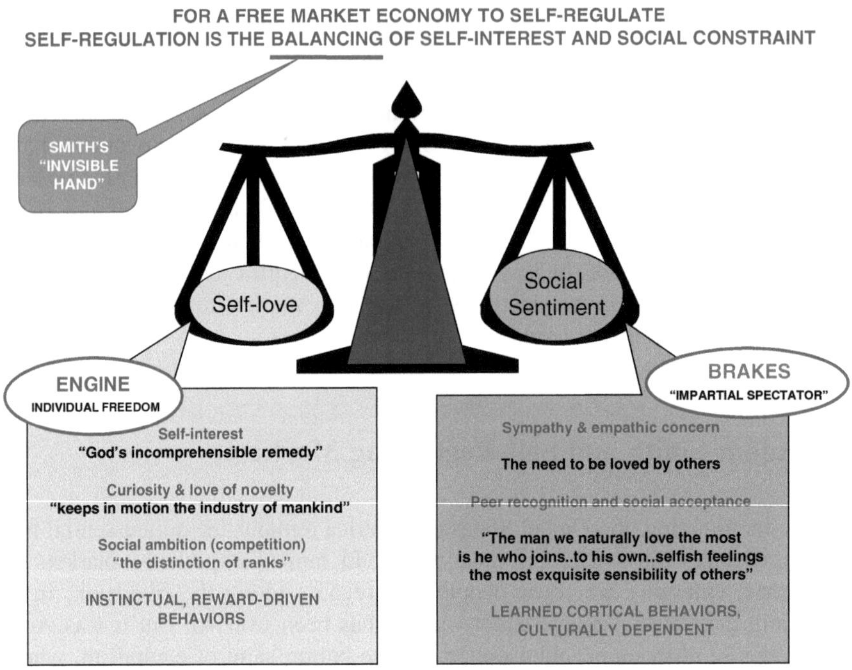

Fig. 2.1 Re-thinking Adam Smith's construct of a free-market self-regulating economy as a dynamic open system

2.4 Changing Cultural Contingencies

For two centuries the United States has pursued Adam Smith's dream of the free-market as the delivery vehicle for universal opulence, with great material success. As the nation of bold ideas, big cars, fast food, sky thrusting cities, and unparalleled military power, America has become a monument to market principles and to the ambition and industry of its people. Whether the old philosopher and founding fathers would recognize themselves as the architects of the competitive, super-charged culture of desire that is the self-image of America today, however, is questionable. Their experience, after all, was of an agrarian and mercantile economy. Adam Smith was writing before the Industrial Revolution. Within the close-knit towns and rural villages of eighteenth-century Britain and of colonial America there was a social intimacy that has almost completely disappeared from American life. Two centuries ago the market systems that Smith championed were embedded in the industry of the local people. Businesses reflected local capital investment and to be solicitous of one's neighbor was prudent insurance against future personal need. Thus the economic goal for most individuals was both private advancement and the social welfare of the community. Through market practice self-interest served the common good.

From the beginning America adopted a competitive commercial thrust, shifting away from grand philosophical visions and toward material gain, first seeking to rival Britain and then to dominate all others as the world's greatest trading nation. Perhaps no nation on earth has more warmly embraced the vision of a technology driven future, and rightfully the twentieth century has been described as the American Century. But over the last three decades, hand in hand with America's market deregulation and growing commercial hegemony, personal opportunity has narrowed as social mobility has declined and the gap between the have and have-nots has widened. As the nation has grown in wealth the founding vision of *novus ordo seclorum*—A New Order of the Ages—has devolved, not necessarily by intent but nonetheless with great consequence, into a scramble for social status and material riches. In the American mind material advantage and human progress have become confused.

Thus today, when the American Dream is magnified through the commercially tinted lens of a globalized, technology-driven consumer culture the neighborly impulse to serve the social good has diminished for such behavior offers little opportunity for reward. The maximization of material wealth is now America's yardstick of social success. In the race to "get ahead" and to triumph as an individual it is competitive struggle and conspicuous consumption that dominate the daily experience. The cultural and economic landscape in which American's live has shifted dramatically and the interplay between social concern and individual desire has shifted with it, disrupting the vital balance that Adam Smith held so dear. The consumer markets in contemporary America have adopted a new social set point—one of maximizing profit.

2.5 Dynamical Systems and Maximization

To adopt a goal of maximizing market profits—as in striving to maximize monetary reward, information, or the food supply—seems reasonable enough but it turns out that when market forces are insufficiently regulated certain problems emerge. Markets have much in common with living organisms in that they each are dynamical systems that seek spontaneous order. Ludwig von Bertalanffy (von Bertalanffy 1969), a father of general systems theory, was one of the first to propose the idea that living creatures avoid entropic disorganization, as would be predicted by the second law of thermodynamics, by maintaining a dynamic equilibrium with their environment through the consumption of energy.

Hayek (1988), the Nobel prize-winning economist who was influenced by Bertalanffy, extended these ideas to argue that an economic system is similar in that it establishes its own extended order. While the market is a result of self-interested human action, Hayek argued, its self-correction does not result from human intention. Rather through the actions of millions of individuals who have the freedom to choose—equivalent in a biological system to, lets say, the individual neurons of the brain—a spontaneous order emerges that has a well-structured, dynamic and self-correcting social pattern. The fundamental and common principle of these dynamical systems, biological and social, is that they are regulated at all levels of their organization by mechanisms that provide continuous homeostatic correction. Thus dynamic systems have the capacity to adapt to changing circumstance although—as is particularly pertinent here—such capacities are not infinite.

The principle control mechanisms of dynamical systems are feedback loops—where raw material, production and product are intimately connected—operating around a set point and designed to sustain an internal environment conducive to self-preservation and competitive survival. Simple examples of such feedback control mechanisms are product price in a market system and in the living system, available energy. Our basic instinctual drives—for sustenance, sex and safety—are controlled by such feedback loops but such dynamic systems have their limitations when driven toward maximization. The set points around which such systems operate to maintain homeostasis can adapt to changing environmental circumstance but ultimately the balance of the system will be compromised if driven to extreme. Maximization—as reflected in an abundance of opportunity for example—thus can distort or disable the necessary regulatory feedback and so disturb the systems capacity to sustain equilibrium. Thus maximizing food intake will rapidly satisfy appetite in the short run but if sustained without regulation the ultimate, and undesirable, long-term outcome will be the toxic state of obesity. Similarly maximizing hedonistic pleasures can lead to addiction.

2.6 The Paradox of Abundance

How complex systems behave can help explain the paradox of abundance—that as choice and material prosperity increase health and personal satisfaction frequently decline. This conundrum highlights a disturbing truth about modernity and human behavior. Having evolved under conditions of danger and privation, we are by instinct a curiosity-driven and pleasure-seeking species focused upon short-term reward. It's a survival mechanism. But, in affluent times, when desire is no longer constrained by limited resources, we have trouble curbing our craving—be that for the fat and sugar of fast food or for the gadgetry of modern technology. This state of affairs comes with little surprise to the behavioral neuroscientist, for it is established that "overloading" the reward circuits of the human brain triggers craving and insatiable desire. In short, the brain's regulatory systems are easily confused by abundance: when it comes to self-indulgence our biology offers no built-in braking system.

We have come to accept that an addict can become habituated to cocaine, heroin or alcohol. But it is the same neural architecture that in a "normal" person promotes habituation to the pleasures of abundance: to the double cheese-burger, to credit-card shopping, to video-games, smart-phones, electronic social networks, the gambling of stock options and to the countless other titillations on offer in the consumption driven society (Whybrow 2009). And there is irony here. In maximizing material choice America has built a market culture that not only reinforces such behavior but also is dependent upon it to function economically. Thus, increasingly, commercial success is measured not by the quality, but by the quantity of product sold—by the merchant's ability to maximize profit. The globalized rich world economies are now dependent upon inducing and sustaining addictive-like behaviors—in America the consumer accounts for some seventy percent of economic activity—and hence the *amount* we consume has become a measure of economic vitality. When portrayed in the media and the glossy magazine this is a world of choice, excitement, energy and self-actualization, but from the perspective of personal health and happiness it is also a world of challenge, mismatch and unintended consequence.

2.7 America's Obesity Epidemic

The growing prevalence of obesity in the US serves to highlight the paradox and the challenge posed by modern-day abundance. While leading the world in material wealth, living standards, freedom of choice and extraordinary technological development, Americans also have the dubious distinction of being among the fattest people on earth. Sixty-eight percent of the US population is overweight and of that group some 33 % are considered to be obese, which to give some perspective is ten times the obesity rate reported for the Japanese. This predisposes millions of Americans to type II diabetes and cardiovascular disease (CDC 2011).

Physiologically the equation is a simple one. A gain in body weight is the direct result of an energy surplus: that over time the calories of energy available from the food consumed by an individual are greater than the energy expended. And so, presuming that it is unlikely that the entire American population has fallen victim to genetic mutation, we must begin to wonder about a mismatch between human behavior and the culture of abundance that we have created. How for example, as a place to start, are the dynamics of the physiological equation influenced by culturally driven life-style changes?

Careful analysis suggests that Americans have been slowly gaining weight for several decades, but there's no doubt that beginning sometime in the 1980s the curve began to rise exponentially (Komlos and Brabec 2010). This timescale corresponds with rapid globalization of the food supply and an increased consumption of energy dense foods containing high levels of sugar and saturated fats, in combination with reduced physical exercise. But other cultural shifts were also in the wind. As Avner Offer, the Emeritus Chichele Professor in Economic History at the University of Oxford, has observed the obesity epidemic corresponds in time not only with the promotion of high density prepared foods but also with the rapid rise of globalized deregulated markets systems that have intensified competition, dramatically increasing the stress and time urgency experienced in the American workplace (Offer et al. 2010).

From the behavioral perspective what is the human impact of the materially rich, information saturated maximized world that we have created? Many of the physical factors that once bridled human behavior have fallen away. Aided and abetted by the convenience of the World Wide Web, instantaneous electronic communication and a revolution in transportation, time and distance are no longer barriers to globalized commercial growth. In this competitive "Fast New World" we have become tethered to the workplace around the clock with time becoming the limiting factor in securing financial and social success (Whybrow 2006). In response to this helter-skelter existence—in our desperate search for more time, more goods, more money—we forgo exercise, rob ourselves of restorative sleep and grab food on the go (Whybrow 2011).

Among those most affected by this cultural shift—and among whom obesity is prevalent—is the hard-working average American, those individuals who toil long hours, often to the neglect of their families, with marginal financial security. In the US the median wage has been stagnant for two decades and there is a growing disparity between the rich and the poor. As was reported by the Congressional Budget Office in 2011 the top one percent of earners in America has more than doubled their share of the national wealth since 1980, now capturing two out of every three dollars of income growth (Pear 2011). In parallel, boosted by the recession that followed the financial crisis of 2008, for the average citizen the competitive frenzy, stress and uncertainty has worsened. Also of significance is that the nature of "work" is changing. For previous generations physical labor was dominant and pre-processed foods were virtually unknown. Today we labor less and eat more. Fast food menus delivering low cost fare high in salt and fat, and soft drinks laced with caffeine and corn syrup, offer a gustatory experience both novel and irresistible to the poorly regulated appetites of the ancient brain. And when we acquiesce to the temptation—not just today, but each day—preoccupied as we are with achieving short-term financial gain and stressed by our exercise poor but treadmill existence then weight gain is not far behind. "Maxing out," as the saying goes, can make you sick (Fig. 2.2).

America's obesity epidemic, therefore, can be understood only within its cultural context—as just one of a cascade of health problems and dysfunctional behaviors that have been triggered by the mismatch between our evolved adaptive biology and the way we live in a rapidly changing cultural environment. It is no accident that in their prevalence obesity, type II diabetes, vascular disease, anxiety and depression are found clustered together for they are each ailments of an affluent, demand driven life-style. While America has been the stalking horse in this behaviorally driven crisis, we are not alone. Such problems are widespread and growing. For example, the World Health Organization has documented in a report published in 2000 that morbid obesity is now a global challenge imposing substantial economic burden and a growing threat to personal health not only for those living in the industrialized nations of Europe but also in the developing world, particularly in the Middle East and China (WHO 2000).

Fig. 2.2 The ailments of affluence: a cascade of health consequences

2.8 Markets and Continuous Growth

In the affluent society maximizing profits in the marketplace necessitates creating need in the mind of the consumer. Fortunately for the merchant this has not been difficult when it comes to foodstuffs. Our primitive affinity for fat, sugar and salt—all elements of diet that were scarce in the ancestral environment—has made it relatively simple for the food industry to stir our ancient cravings. Such appetites are reward driven and innate, wired deeply into the survival mechanisms of the primitive brain. Thus when consistently reinforced desire can rapidly run away to greed, as is evidenced by the obesity epidemic. In general, however, creating a consumer culture of continuous growth, with the goal of maximizing profit, is more of a challenge: inherently maximization and the self-correcting dynamics of the marketplace are incompatible. This is a cardinal lesson to be learned from the Great American Experiment.

In theory the beauty of the market society is that no one individual controls its growth simply because the "price" of any barter—the value placed upon any exchange—must be agreed upon by the parties involved. In any transaction when it comes to decision making we intuitively evaluate opportunity against risk: we balance the possibility of personal reward against fear. Commonly this evaluation is made with the benefit of experience. We do not run into the street in the face of oncoming traffic to pick up a dollar bill that we find blowing in the wind: nor in the market place do we enter into partnership and put our resources at risk with individuals whom we distrust. We avoid impulsive decisions under circumstances that we do not completely understand: indeed, not to do so is called foolishness. We accept that the world is not perfect; that risk exists and that we will not always get what we want. So we curb our enthusiasm when necessary. In short we exercise prudence in our affairs, including in the market place.

In the interests of maximizing profit the consumer society, of which the US is the exemplar, seeks to stand this reality on its head. And indeed, beginning in the 1920s, and accelerating in the last decades of the twentieth century, this common sense practice of barter has been eroded. Slowly, as trade has become globalized, markets have become divorced from the reality of everyday experience, especially in the financial industry where money has replaced labor in defining wealth. Finance we are told is a world apart, the necessary lubricant of the world's economic health, and distinct from the mundane propensities that explain human action in the market place. The global financial crisis of 2008 reminds us that this is a falsehood—that dynamic systems, forever seeking balance, create their own course correction.

In the electronic age money as a tangible asset has become invisible, just a string of numbers on a computer screen recording the work that we do and the expenses that we incur. Silver and gold are no longer associated with money: even paper money is going out of style replaced by plastic cards of credit. And yet money has become ever more central to our lives. We equate it with "the market". It's the accumulation of those numbers that makes us rich or poor. For some it is the string of numbers on the computer screen that provides identity. Our belief in the future is tied increasingly to their continuous escalation and thus in the public mind abstract financial markets have become the main index of economic activity.

The danger of such abstraction is that it removes us from the realities of everyday life—of managing one's own affairs within familiar parameters. Thus there is a consensus that the financial wizardry and loose credit that spurred the American housing bubble and the financial crisis of 2008 did so largely because in the pursuit of maximizing profit several leading banking houses indulged in excessive speculation. Bundling loans together in ways that bore little relationship to reality ultimately, and inevitably, drove the financial markets to a course correction and the speculative bubble burst with worldwide consequence.

The desire to own one's own home has long been a cornerstone of the American Dream. New research from the University of Chicago business school, however, using data collected by US government agencies from three thousand zip-code districts suggests that the explosion of mortgage growth in America that preceded the housing bubble was tied to easy credit, particularly in those areas of the nation least able to afford it. Startlingly it was those districts where households frequently had been rejected for mortgages prior to 1996 that had the highest rate of approved mortgages in the boom years between 2001 and 2005. These districts termed "high latent demand" zip codes by the lending institutions were regions of declining income and poor employment growth during those same years (Chain of fools 2008, p. 84).

In other words in the hope of turning a quick profit as housing prices escalated the risk standards applied to lending to these neighborhoods had been relaxed making it highly probable that with a downturn in the economy individuals would begin defaulting on their loans. The risks were high and yet these new loans were bundled with other mortgages and sold to unsuspecting investors. With such borrowing practices new dynamics entered the equation: not only was the lender insecure but also the mechanisms to insure against risk were later determined to be suspect. In addition, personal risk on the part of the individual bankers and insurance agents was diminished further because the money financing the loans was derived from public offerings rather than from the capital reserves of the lending bank. Thus the resulting derivatives—items designed to minimize risk—were no longer firmly grounded in reality and the dynamics of the market became distorted. With hindsight the lesson is clear. Just as in biology, where the maximization of one particular cell type is disruptive and called a cancer, attempts to engineer the maximization of profit in a dynamic market economy similarly can end in disorder and disaster.

2.9 Lessons Learned

The American Experiment has generated great material wealth but it has also provided some important insights into human behavior. Inadvertently, among other lessons, we have stumbled upon a new behavioral maxim: that the better human society becomes at providing instant gratification then the less capable each individual citizen becomes at self-regulation. It is a curse that we have visited upon ourselves, for it is the very abundance of American society—we produce more, consume more, and throw away more than any other people on the planet—that nurtures our consumptive greed. In America, and increasingly as consumerism spreads across the rest of the world, we are becoming addicted to novel, compelling enticements of our own manufacture.

But the Great Experiment also offers another lesson. In the eighteenth century the concept of happiness was inextricably linked with the effort to create a science of man, one that equated desire, personal ability and reward within a dynamic construct. When desire outran the ability to satisfy it, then misery could be expected. Faced with such circumstance the common sense approach was to decrease one's desire, to increase one's productive engagement or preferably to do both. The importance of "common sense" was invoked: this was a quality not found in the individual alone but rather in the shared wisdom of the community, a communal sense, as reflected in Thomas Jefferson's *Commonwealth* of Virginia. Such sensibility drew upon an intuitive body of truth vouched for by experience and common suffrage. These were the "self-evident truths" of Enlightenment philosophy—the pursuit of life, liberty and happiness—that Jefferson deemed worthy of citation in the *Declaration of Independence*. For modern-day nations, including the US itself, it is an observation that also remains worthy of reflection. Indeed it is perhaps the most important lesson that we have learned to date from the Great Experiment: that the affluence we have worked so hard to maximize has the potential not only to be *constructive* but, if misappropriated or poorly harnessed, also to be *destructive* of a society's health and happiness.

References

Center for Disease Control and Prevention (CDC) (2011) Overweight and obesity: data and statistics. National obesity trends, 1985 to 2010. Available at: www.cdc.gov/obesity/data/trends.html. Accessed on 1 Dec 2011

Chain of fools: Hard evidence that securitisation encouraged lax mortgage lending in America. (2008) The Economist, 9 Feb, 84

Congressional Budget Office (CBO) (2011) Trends in the distribution of household income between 1979 and 2007. Available at: www.cbo.gov/publications/42729. Accessed on 1 Dec 2011

De Tocqueville A (1835) Democracy in America. Anchor Books, New York

Hayek F (1988) The fatal conceit: the errors of socialism. University of Chicago Press, Chicago

Kessler RC, Berglund PA, Demler O, Jin R, Merikangas KR, Walters EE (2005) Lifetime prevalence and age-of-onset distributions of DSM-IV disorders in the national comorbidity survey replication (NCS-R). Arch Gen Psychiatry 6:593–602

Komlos J, Brabec M (2010) The trend of mean BMI values of US adult birth cohorts 1882–1986 indicates that the obesity epidemic began earlier than hitherto thought. Am J Hum Biol 22(5):631–638

Lemay L, Zall PM (eds) (1986) Benjamin Franklin: autobiography. WW Norton Critical Editions, New York

Meritocracy in America (2005) The Economist, 1 Jan, 22–24

Offer A, Pechey R, Ulijaszek S (2010) Obesity under affluence varies by welfare regimes: the effect of fast food, insecurity and inequality. Econ Hum Biol 8(3):297–308

Pear R (2011) Top earners doubled share of nation's income, study shows. New York Times, 25 Oct

Smith A (1776) An inquiry into the nature and causes of the wealth of nations. Methuen & Co, London

Upper bound: Social mobility and inequality (2010) The Economist, 15 Apr, 37–38

von Bertalanffy L (1969) General systems theory. George Braziller Inc., New York

Whybrow PC (2006) American mania: when more is not enough. WW Norton and Company, New York

Whybrow PC (2009) Dangerously addictive: why we are biologically ill-suited to the riches of modern America. The Chronicle of Higher Education

Whybrow PC (2011) Time urgency, sleep loss and obesity. Proc Br Acad 174:129–140

Wills G (1978) Inventing America: Jefferson's declaration of independence. Doubleday and Co, New York

World Health Organization (WHO) (2000) Obesity: preventing and managing the global epidemic. WHO Obesity Technical Report Series, 894. World Health Organization, Geneva

Chapter 3
More Nonsense and Less Happiness: The Uninteded Effects of Artificial Competitions

Mathias Binswanger

3.1 Introduction

In many sectors of the economy, functioning markets do not exist. Therefore, due to an increasing faith in markets and competition over the past few decades, there have been more and more attempts to induce artificial competitions in sectors such as health care, education or science in order to promote efficiency, where markets are absent. But these artificially induced competitions do not enhance efficiency. Instead they lead to the production of nonsense. This can be explained as follows. On a functioning market, producers of goods and services have an incentive to meet the needs of consumers, as this maximizes their profits. But in artificial competitions without markets people's needs do not matter. Instead, these competitions induce individuals to concentrate on measurable performance indicators.

Thus, a new spectre is haunting Europe. It is the spectre of artificial competition, which has developed into a new ideology. For example in science, professors and universities are ranked according to the number of publications in scientific journals as an indicator of scientific excellence. These artificial competitions incentivize scientists to maximize the number of publications while the content of these publications becomes increasingly irrelevant. Therefore, the resulting "excellence" is in fact frequently nonsense.[1] The artificial competition about

[1] This statement does not imply that competitions are generally harmful or negative. Competitions are as old as the history of mankind and play a central role in most societies. Adam Smith himself wrote that the drive to compete with others appears inherent to the human condition (Smith 1759).

M. Binswanger (✉)
Institute for Competitiveness and Communication, School of Business of the University of Applied Sciences Northwestern Switzerland, Olten, Switzerland
e-mail: mathias.binswanger@fhnw.ch

H. Brockmann and J. Delhey (eds.), *Human Happiness and the Pursuit of Maximization*, Happiness Studies Book Series, DOI: 10.1007/978-94-007-6609-9_3,

"winning the publication game" misleads scientists to concentrate on the number of publications instead of on the quality of their research.

In Europe artificial competitions were mostly advanced by socialist governments over the last decades. They followed the politics of their new leader Tony Blair, the former prime minister of the United Kingdom, who inherited the enthusiasm for competitions from his conservative predecessor Margaret Thatcher. His enthusiasm was based on the fact that the use of artificial competitions gave socialists a new argument in favor of government. The old-bureaucratic government was supposed to be replaced by a new efficient government where new public management would rule. And new public management heavily relied on artificially induced competitions and so-called pseudo-markets. Therefore, these competitions became part of the standard program of "progressive socialists". But, not only politicians, but also consulting firms like McKinsey jumped on the bandwagon. It gave them a new opportunity to "advise" government institutions such as universities or hospitals with the same concepts and tools as private companies. And they could preach competition as a general remedy for increasing efficiency. Thus today, from right to left across the political spectrum, competition is propagated without being aware of its negative consequences in a non-market environment.

3.2 Artificial Competitions Without Markets

Often the terms "market" and "competition" are considered to be Siamese-twins. We tend to think that wherever there is a market, there is also competition and vice versa. However, this is not true. On the one hand, we have markets with very little competition, such as in the case of a monopoly or a trust. And on the other hand there are lots of competitions without any relation to a market such as sports competitions. In competitions outside a market environment, there is no price mechanism that induces an adjustment of supply to demand, as is the case for competition in a functioning market.[2] Instead, runners at the Olympics compete for medals, scientists compete for research projects, and knights used to compete for the favor of a lady. All of these are examples of competitions outside a market environment.

However, there is a commonly shared view that competitions ensure optimal results, even without markets. This, for example, is the general thrust of a brochure published by the Swiss megabank UBS, titled "Management of Public Organizations" (2005, p. 20), where we find the following statement (translated by the author):

[2] According to Max Weber, competition is simply described as the pursuit of a goal by at least two individuals. For more general information on competition and the various forms in which it appears, see Nullmeier (2000a).

> The government must initiate competition in all areas, where tasks or the supply of services cannot be left to the free market... Wherever and for whatever reason a market for the supply of public services cannot be established, competition-like measures have to be introduced instead.

Following this kind of thinking, there have been lots of attempts to induce artificial competitions in domains such as science, education, or health care, where leaving the supply of goods or services to the market would lead to undesirable results. Therefore, as in sports, a never-ending contest for the best performance is supposed to provide the same beneficial effects as the price system on a functioning market. In reality, however, this idea proves to be wishful thinking. If "competitions without markets" was a promising idea, communist command economies would have worked quite well. These economies had no markets, but where full of artificially induced competitions, in order to create incentives for working more efficiently. For example, in the former German Democratic Republic these competitions were called "socialist competition". Lenin himself wrote after the successful revolution in Russia: "Now, since a socialist government is in power, our task is to organize competition" (Lenin 1961, p. 405). However, the communist command economies with their artificially induced competitions failed miserably, and the same is also true for our current attempts to induce artificial competitions.

An example of the times of command economies illustrates the problem quite well. Ota Sik, the former Czech minister of economic affairs at the time of the Prague-Spring-movement in 1968, who later became a professor of economics at the University of St. Gallen in Switzerland and whose lectures I attended in the 1980s, told us the following story. As many other industries, the shoe industry in the Soviet Union was characterized by low productivity and a tremendous waste of resources. No one had an incentive to make much effort, since wages were low and completely independent of the level of individual effort. What was there to do in this situation? The most obvious solution, the introduction of markets, was not possible for ideological reasons. So it seemed that artificially induced competitions would be the only remedy. Thus, economic experts began to search for performance indicators, which would measure workers' efforts in the shoe industry.[3]

The experts came to the "brilliant" idea to use "material use" as the relevant performance indicator and to pay the "best" worker collectives, who used up most material, corresponding "performance bonuses". The idea behind the use of this indicator is easy to understand. Whoever produces more shoes needs also more material, which can be measured by units of weight. Therefore, there should be a positive correlation between "material use" and "work efforts". However, the result was different from what the experts had imagined. In the course of a few years, shoes became heavier and heavier. The previously hardly motivated workers started to become innovative and continually developed more material-intensive

[3] See also (Rosen 1988, p. 81) and the artificially induced competitions about work points, with were used in communist China.

shoe models. Material-intensity is, however, not a feature, which consumers are really looking for when buying a new shoe. Instead of making the shoe industry more efficient, the artificially induced competition resulted in weightier and less comfortable shoes, which finally no one wanted to wear. The heavier "competitiveness" of the Soviet workers in the shoe industry proved to be fatal.

At hindsight this example looks highly absurd and we may consider the episode as just another proof of the inefficiency of a command economy, which however belongs to the past. Indeed, the production of shoes functions well in current market economies and the customer can choose among an enormous supply of stylish and light shoes. However, when we look around more carefully, we can observe quite similar phenomena today as in the former Soviet shoe industry. Once more, thousands of employees meticulously produce services and products, which no one really needs or wants. However, the nonsense is less obvious as it was in the former Soviet shoe industry. There, people were confronted with the nonsense in their daily life, whenever they wanted to buy new shoes. But the majority of people are only indirectly confronted by the nonsense resulting from today's artificial competitions in science, education, or health care. However, there are still very obvious cases even today, as the following episode from modern England illustrates.

The Neue Zürcher Zeitung from December 8, 2007 reported that at the beginning of the new millennium, British city administrations began to outsource various public services in the course of the privatization mania of those years. This also affected the management of public parking lots, where issuing parking tickets is a major task. Why should the government issue parking tickets, if private enterprises can also do this job? In order to give the privately hired parking attendants an incentive to really search for illegally parked cars, an artificial competition was induced as well. The payment of the parking attendants was made dependent on the number of parking tickets they issued. And the hardest-working parking attendants received televisions or even automobiles as bonuses.

Like the workers in the Soviet shoe industry, the parking attendants became very "efficient" and "innovative" due to this artificial competition. They were eagerly looking for expired parking meters, in order to issue as many tickets as possible. As British newspapers reported, parking tickets were issued even before the drivers had a chance to place their coins into the meters. Other parking attendants began to manipulate photos, which were used as evidence, and a bus even received a ticket for stopping at a normal bus stop. This over-enthusiasm caused by an artificial competition on the part of the parking attendants, finally provoked such massive protest that the parliamentary committee on traffic was forced to deal with the matter in 2006. However, instead of fighting the cause they decided to fight the symptoms, since otherwise they would have had to put the entire policy of privatization into question. They attempted to organize a "fairer" competition, by creating a new code of conduct for parking attendants. But the incentive to issue as many parking tickets as possible was not removed.

As is shown in these examples from the former Soviet Union and from modern England, competitions without a market generally do not lead to an increase in welfare. If an incentive to consume as much material as possible is set, workers will behave accordingly. And the same is true for the incentive to issue as many parking tickets as possible. But all of these efforts do not correspond to people's needs. They neither want shoes with a maximum weight, nor parking lots, where parking attendants are obsessed by the idea of issuing as many parking tickets as possible.

These examples illustrate the major problem of artificially induced competitions outside a market environment. The true needs of consumers of a good or a service have no influence on the efforts of suppliers of these goods or services, which instead are directed by performance indicators. These performance indicators are set by government authorities or managers and are supposed to measure effort and quality of output. But there is no market price system, which in a functioning market ensures, that supply meets demand as closely as possible. If a shoe market had existed in the former Soviet Union the price for the heavy shoes would have quickly declined to zero due to a lack of demand for these shoes. And suppliers would surely have started to produce lighter shoes, which consumers also wanted to wear. For this reason, we currently leave the supply of shoes to the market, since this leads to a much more efficient result as compared to the result of an artificially induced competition in a command economy.

In the case of parking tickets, on the other hand, there is no market solution. Here we deal with the "production" of a public good (maintaining order in public parking spaces), and there is no incentive to provide public goods in a market economy. However, the government can let parking attendants fulfill their task without an artificially induced competition. This will lead to a better solution, where parking attendants are not incentivized to become a public menace.

3.3 Performance Measurement: 100 m Run Versus Figure Skating

Most of the relevant efforts in real life cannot be measured by numbers, since they deal above all with quality and not quantity. And all the attempts to measure quality with the help of quantitatively measurable performance indicators lead to perverse exclusively incentives. A short excursion into the world of sports can help to illustrate this phenomenon. In many sports disciplines performance measurement is rather easy. Under the condition that an appropriate measurement technique exists, one can easily determine how long an individual runner needs for running 100 m or at what height a high jumper crosses the bar. In these cases we are dealing with performances that can exactly be measured on a cardinal scale. In other words, one can specify by how many time units one runner has been faster than another, making it possible to create an exact ranking of the performance of all runners. The objective is to run as fast as possible, which results in a competition, where runners try to outperform each other when running the 100 m. The goal of the competition and the motivation of the individual runners perfectly match in this case.

The relevant performances in education, science or health care, however, are much more difficult to measure than the time in a 100 m run. The problem becomes evident once we look at more complex sport disciplines like gymnastics or figure skating. In those disciplines also the quality of a presentation is relevant for the final ranking. But this quality cannot be measured, and therefore the performance of two athletes cannot be compared on a cardinal scale. We can no longer say, for example, that the Olympic champion in figure skating was 1.5 times as "artisty" as the one gaining the bronze medal. But since in sports rankings of all participants are needed, the quality of a performance somehow has to be measured. In this case the measurement is done with the help of a jury of experts, who seems to be very able to this job. Only a few seconds after the performance of a figure skater is over, the members of the jury come up with a precise number, which is supposed to reflect the quality of the performance.

However, in reality, even a jury of experts is not able to measure quality. Instead, the expert juries base their evaluation on quantitatively measurable numbers, which then must serve as indicators for quality. In figure skating, the major indicator is the number of successfully accomplished triple, or even quadruple jumps. Thus, figure skating has developed into a rather strange discipline, where as many complicated jumps (e.g., Axels, Salchows, or Rittbergers) as possible are presented during the limited time of a free program. These jumps are supposed to impress the expert jury, since the skaters know that the assessment of the quality will be based on this measurable performance indicator. A more accurate name for this sports discipline, therefore would be "musical ice jumping".

Sports disciplines such as figure skating show the impossibility of directly measuring and comparing the quality of performances. Measurement relies on numbers, which hopefully correlate with the quality of a performance. In the example of figure skating, the number of successfully performed multiple jumps, which additionally are somehow weighted according to their degree of difficulty, serves as an indicator of quality. This kind of "quality measurement" assumes that the quality of a free program in figure skating is the higher the more complicated jumps are presented. By this approach, however, quality is reduced to a measurable indicator. And in this case the actual goal of the competition, namely to provide high-quality performances, is no longer identical with the motivation of an individual skater. Instead, she is incentivized to present as many complicated jumps as possible and this goal replaces the original goal of quality.

Of course, in the case of figure skating, the crowding out of the non-measurable quality by measurable jumps is not a big deal. As long as figure skating entertains the audience, it does not matter how the quality of a free program is really measured. Top performances in sports must fulfill no other goal than that of entertainment, and the resulting increase in triple and quadruple jumps does not harm anybody even though these jumps do not really make sense. In the real world, however, this problem is not innocuous. If the relevant quality of a performance is crowded out because the participants of a competition focus on measurable indicators, it has negative consequences for the well-being of those who are affected by the crowding out of quality (see also Smith 1995, p. 284).

3.4 Qualitative Performances and the Futile Attempts to Measure Them

Today's relevant performances have much more in common with figure skating than with a 100 m-run, where quality does not matter. Of course, we can always define measurable indicators such as the number of multiple jumps in figure skating. But one cannot capture the real quality of a performance with such methods, even if more and more measurable indicators are introduced. The more exactly we attempt to measure specific aspects of quality, the more we start to neglect other, less measurable aspects.

The scientists and consultants who claim to "measure quality" are in a similar position as quantum physicists. The more exactly they attempt to determine the position of a particle, the less exactly they can measure its momentum and vice versa. It is well-known that a so-called Laplacean demon cannot exist, who by an increasingly exact measurement of position and momentum of all particles of a system, would gain more and more information about this system. The famous French physicist Laplace postulated this demon in the 19th century, when physicists still believed in the mechanistic paradigm of Newtonian physics. However, as physicists later found out, the more exactly we try to describe a system by measuring all of its properties, the more, the system is changed by the measurement process itself. If one attempts to determine the position of particles with increasing precision, more and more energy is needed to do so, which in turn will change the position of the particles, which was supposed to be measured more precisely.

For a social system such as a private or a public organization this is even truer. Attempts to measure quality of certain performances more precisely, will change the system itself (private or public organization) as people, who are affected by these performance measurements, will no longer behave in the same way as before. In spite of this knowledge, scientists and consultants working in the field of quality measurement still seem to believe in the Laplacean demon and, therefore, in the exact measurability of qualitative performances. They tell us that we only have to use more and better indicators as well as more and better computers. But this kind of thinking is epistemologically outdated and brings us back to the 19th century, when the illusion of the Laplacean demon was still alive.

The difficulty of measuring quality of work performance has also become increasingly obvious in the course of economic development. When industrial production still dominated in many countries, measuring workers' performance was relatively easy. An assembly-line worker could be assessed either by the number of hours he worked or by counting the number of processed items. Both indicators possessed a high correlation with the relevant performance. Once an assembly line started to run, workers had no other option than working, as nicely illustrated by the film "Modern Times" by Charlie Chaplin. In this case, the measurement of the performance was still comparable to a 100 m run, since qualitative aspects were not important. But fortunately such kind of work has become increasingly rare over the last decades.

However, attempts to measure only slightly more complex activities, such as secretarial paperwork in an office, immediately led to absurd results. The Lincoln Electronic company, for example, tried to measure the performance of its secretaries by the number of characters, which they typed on their typewriters. The assumption was that the more letters are typed the better is the performance of a secretary. But soon this heroic attempt of performance measurement had to abandoned again, since secretaries were constantly hitting the keys of their typewriters also during breaks (Fast and Berg 1975). In this case, the measured improvement in the value of the indicator did not correspond to an increase in the relevant performance, but rather led to an increase in a meaningless production of typed characters.

Even more problematic are the attempts to measure performance related to genuinely complex and creative activities. If the performance of people working as scientists or doctors is measured in the same way as the performance of assembly-line workers, the measured performance will no longer correspond to the desired performance. Innovative ideas have often been developed in only a few minutes or in only a few pages, while on the other hand, long hours and thousands of articles and reports did not lead to useful results (see also Lotter 2008, p. 54). If the performance of a researcher, for example, is measured by the number of articles published in scientific journals or the number of projects he has completed, it may measure his diligence, but certainly not his relevant performance.

3.5 How Artificially Induced Competitions Create Perverse Incentives

In artificially induced competitions, there is usually not much correlation between measured performances and actually relevant performances. As an example, let us take customer satisfaction, which is an important goal in many organizations. Today everybody seems to be a customer. Citizens are customers of public administrations, students are customers of universities, and patients are customers of doctors or hospitals. However, "customer satisfaction" cannot directly be measured. Thus, measurable performance indicators are used, which are supposed to correlate with the unmeasurable "customer satisfaction".

Scientists and consultants have been busy with the development of such indicators for a long time. One aspect of customer satisfaction is the fast and efficient handling of complaints. Therefore, we may collect data about the percentage of customers, who had to wait longer than 10 days for their complaint to be handled. This indicator has the advantage that it is easy to measure, but the disadvantage that it already creates a perverse incentive. If employees' performance is measured by this indicator, they will concentrate their efforts on taking care of complaints from customers that have waited for 8 or 9 days, so that the limit of 10 days is never exceeded. However, they do not improve the measured performance if they process complaints that have just arrived. Therefore the average time for processing complaints may even increase, which at the end does not promote but rather harm customer satisfaction.

Naturally, the responsible managers would sooner or later notice this counterproductive result from customer reactions, and come to the conclusion that "the number of complaints that have not been processed for over 10 days" should not be used as a performance indicator. An alternative indicator would be "the average time required for handling customer complaints". However, the delight of having found this solution will also be short-lived. If employees are confronted with the new indicator, they will have an incentive to concentrate their time on dealing with the easy cases and answer them as soon as possible. Difficult cases, on the other hand, will be neglected, since processing them will no longer "pay off". In this way, the measured indicator will improve, but the customer with unanswered complaints will hardly keep their discontent to themselves and this will sooner or later negatively affect the reputation of the organization.

Both of the behavioral reactions, which just have been described, are rather typical. In the first case, employees concentrate on fulfilling a specific measurable aspect of customer satisfaction and forget the actual goal of their activity. This phenomenon has also been described as "measure fixation" (Smith 1995, p. 284) and occurs most often when complex performances are assessed with an easily measurable indicator. This fact has been known for a long time as "Campbell's Law", which is derived from the work of the social scientist Donald Campbell. In 1976 he wrote (p. 49):

> The more any quantitative social indicator is used for social decision-making, the more subject it will be to corruption pressures and the more apt it will be to distort and corrupt the social processes it is intended to monitor.

In the second case, employees concentrate on processing easy complaints and neglect the complicated ones. In doing so, they are "cherry picking", which sometimes is also referred to as "cream skimming" (see, for example, Bartlett and Le Grand 1993, pp. 31–34). Cherry picking becomes a problem when indicators measure the successful completion of specific activities. Another example may illustrate this point even more clearly. If job placement officers' performance is measured by the number of successfully re-employed job seekers, they score well if they concentrate on the easily re-employable job seekers and neglect the difficult cases (see van Thiel and Leeuw 2002, p. 272).

But let's return to our example and the search for indicators to measure "customer satisfaction". Once experts agree that the successful handling of complaints cannot be adequately measured with one indicator, they will probably come to the conclusion that it requires an entire system of indicators. In the simplest case, both of the previously mentioned indicators could be combined. Thus, "the number of complaints that have not been handled for over 10 days" is combined with "the average handling time", and both indicators will have a weight of 50 % in the performance assessment. However, even with such a simple indicator system, the incentives set by this system are not obvious any more. Though the negative incentives set by the individual indicators are mitigated, the same also applies for the intended positive incentives. The incentive to neglect difficult cases will be less pronounced, but the same is true for the intended incentive to reduce the average handling time.

The more complex systems of indicators are introduced, the more complex become the incentives set by these systems. Therefore, it will be less and less transparent how to behave in an optimal fashion in order to score high, and the relation between the performance indicators and the actually desired performance (in this case customer satisfaction) turns into a black box. For example, the performance of British physicians is measured by a system of 146 indicators (Campbell 2009). But systems of hundreds of indicators are just black boxes where the relation between the performance indicators and the desired quality gets lost. Moreover, the development, implementation and evaluation of a large number of indicators also lead to additional bureaucracy. Potemkin villages are built, in which with the help of computers, tons of irrelevant data is collected, processed and evaluated, providing the illusion of gaining increasing knowledge about the quality of performances. But this new bureaucracy is hard to identify, since it is disguised behind slogans, which advertise these efforts as new efficiency tools.

There is also a growing body of literature investigating the behavioral reactions of people to the introduction of performance indicators (see, for example, the cited literature in van Thiel and Leeuw 2002). This literature shows that people are very creative in finding ways how to score high in artificially induced competitions without increasing the true quality of a performance. However, such unwanted effects are always treated as exceptions and not as the rule. Many experts still want to make us believe that these negative effects can be removed by improving the indicators, or by adding more indicators, which brings us back to the fiction of the Laplacean demon. In reality, perverse incentives, which consequently lead to perverse behavior, are the rule and not an exception in artificial competitions.

Perverse incentives are the more pronounced, the less performance indicators correlate with the desired outcome of a competition. Low correlation creates the opportunity to increase the value of indicators without increasing the relevant outcome or even at the cost of decreasing it. This fact is also known as the "performance paradox" (Meyer and Gupta 1994), which is illustrated by the examples mentioned in Sect. 3.2. There, in the Soviet shoe industry the indicator "material consumption" could easily be increased without improving the quality of the shoes. And the increase in issued parking tickets did not lead to better public service on London's parking lots. Indeed, in both cases even a negative correlation between the measurable indicator (e.g., material use) and the quality of the performance (e.g., quality of a shoe) could be observed.

3.6 Competition Becomes an End in Itself

"After the game is before the game" is a famous quote of the former German soccer trainer Sepp Herberger. This also applies for competitions in general, where winning is the first priority. As soon as the winner of a competition is known, the competition is over, and the next competition must start as otherwise everything comes to a halt. Records in sports are there to be broken again, and participants are never allowed to be satisfied with the status quo. Thus there is a constant strive for "better" performances, as this increases the chances of winning the next competition.

Artificially induced competitions lead to a permanent state of rivalry (Rosa 2006, p. 94). The individual is subject to constant pressure to perform, although the performance itself is often completely irrelevant to the outside world. Indeed, once a competition has started, it develops its own dynamics, which affect all participants. Therefore, competitions often result in misguided and excessive use of time or energy, which negatively affects subjective well-being of those being forced to participate in these competitions (see Binswanger 2006, pp. 63–65). The following article from the German *Frankfurter Allgemeine Zeitung* from December 28, 2007 shows this kind of thinking very clearly (Leber 2007, translated by the author):

> Global competition forces us to identify the best talents and give them an optimal education. By doing so, we put them in an international competition for the best performance. Avoiding this challenge would result in economic stagnation. There are no more niches—neither in the market of services, nor in the market of talents.

Striving for competitiveness becomes something like a natural law according to the author of this article.

Often the claim is made that consumers are the real force behind this development. We also find this claim in the already cited article:

> Finally, it is the consumer who wants good services at low prices from a global supply. Wherever the customer, supported by product reviews and the Internet, searches for the best offer at the best price, he demands the brutal competition for the selection of the best products and the top talents.

However, this is precisely not true for competitions without markets. There, consumers are completely absent and do not demand any of the efforts which are stimulated by artificial competitions. Instead, these competitions result in unnecessary stress for all those who are affected.

Nevertheless, performing high in artificially induced competitions seems to become the dominant goal of more and more activities of scientists, students, doctors or professors. Due to this obsession with competitions, the actual purpose of their activities often falls into oblivion and performing high in the competition becomes an end in itself. The politics of research is a good example for this. In 1985, before the ideology of competition really hit science, the German Council of Scientists explicitly declared: "Competition is not an end in itself" (Nullmeier 2000b, p. 213). However, in the meantime this warning has long been forgotten and the primary goal for German scientists is to improve the competitiveness of the German universities and research facilities.

When competitions become an end in itself, the content of these competitions tends to become less and less important. "What is produced, what is researched, is irrelevant at the end, as long as it increases competitiveness" (Rosa 2006 p. 95). What matters is the number of published articles and where they have been published. The content, however, is usually not of much relevance. With the exception of the referees (sometimes not even them), who have to evaluate the articles and decide whether they should be published in a particular journal, nobody will ever read a large portion of the articles published in scientific journals (Binswanger 2010, Chap. 7). Similarly, it does not matter what is exactly studied and learned at a particular educational institution. The main goal is to get a diploma since this is what counts at the end. Therefore, many important aspects of quality are neglected, as they do not directly matter in the artificially induced competitions. But this negatively affects the overall quality of work in many areas including research, education and health care.

3.7 Concluding Remarks: More Nonsense and Less Satisfaction at Work

Artificially induced competitions, in contrast to competitions on functioning markets, do not lead to better and more efficient performances. Only if competition leads to price adjustments, it creates an incentive to adjust supply to demand and, therefore, to people's needs. In this case, the "invisible hand of the market", which was first described by Adam Smith, ensures efficiency, as long as certain additional conditions are fulfilled. However, in competitions without markets the invisible hand is absent. Instead of adjusting supply to people's needs, producers try to get high scores in measurable performance indicators. Therefore, artificially induced competitions create perverse incentives.

Consequently artificially induced competitions also lead to production of nonsense, which is not needed by anybody. Scientists write articles and fill thousands of pages in scientific journals in order to discuss topics, which nobody is interested in. But they want to score high in the publication competition and, therefore, any topic is fine, as long as it will result in another publication More and more young people are educated for more and more years as students of universities, in order to get Bachelor's and Master's degrees for studies, which do not provide them relevant knowledge for their future life. An increasing number of medical examinations and tests are conducted in order to prevent illnesses that never would have occurred anyway. And when we want to select a yogurt or a university that is appropriate for our needs, we are confronted with a bulk of quality labels and certificates, which rarely help us to make a good selection.

These developments, however, we are told, are important to our prosperity and to our future well-being. The more scientific articles are published, the more reforms are carried out, the larger the percentage of students at universities, the more medical examinations we have, the more quality labels are created, the better we are off. But these are naive illusions. The production of useless efforts and services creates jobs, but at the same time, it keeps us from concentrating on those efforts and services which truly matter for our well-being. Sense is crowded out by nonsense, quality is crowded out by quantity, and intrinsic motivation is crowded out by rewards and punishments (see, for example, Deci et al. 1999).

A policy, where people are rewarded for performing well in artificially induced competitions, and where they are punished for performing badly, is essentially a system of "carrots and sticks." In such a system, all employees are put under suspicion to be lazy. It is assumed that they are able to work more and better, if they only wanted to. Therefore, carrots and sticks are necessary in order to "motivate" employees to increase their work efforts. The problem of such a system was well described by Reinhard Sprenger in his famous book "Mythos Motivation", first published in 1992 (Sprenger 2002, p. 42 ff). There we read the important sentence that "motivation is methodized mistrust". Sticks and carrots result in cultures of mistrust, which successfully crowd out intrinsic motivation and, therefore, the joy of working. But intrinsic motivation is of vital importance in areas such as science, health care, or education. Only intrinsically motivated scientists enable scientific progress and only intrinsically motivated doctors provide a truly good treatment for their patients. But intrinsically motivated scientists or doctors are typically people, who do not feel like participating in artificial competitions and working just for getting high scores in measurable performance indicators. If they are managed by "sticks and carrots" they lose a great deal of their motivation or even stop working in such an environment. But since quality is closely linked to intrinsic motivation, the crowding out of intrinsically motivated people also results in a crowding out of quality.

In science, education, and health care, top performances are achieved, if talented and motivated people are given the chance to be creative in a free and stimulating environment. For this reason it is counterproductive to put scientists, professors, teachers and doctors under general suspicion of being lazy, and to assume that they are "black sheep", who will only deliver a good performance if they are rewarded by carrots or punished by sticks. This attitude crowds out the intrinsic motivation of talented and motivated people, which is the reason, why relevant top performances become increasingly rare. And, on the other hand, artificially induced competitions cannot turn unmotivated and/or less capable people into top performers no matter how sweet the reward of a tasty carrot or how frightening the punishment by a stick may be. What they will produce under these circumstances is measurable nonsense, which we don't need.

References

Bartlett W, Le Grand J (1993) The theory of quasi-markets. In: Le Grand J, Bartlett W (eds) Quasi-markets and social policy. Macmillan Press, London, pp 13–34

Binswanger M (2006) Die Tretmühlen des Glücks. Herder Verlag, Freiburg

Binswanger M (2010) Sinnlose Wettbewerbe. Warum wir immer mehr Unsinn produzieren. Herder Verlag, Freiburg

Campbell DT (1976) Assessing the impact of planned social change. Occasional papers series, paper #8. Public Affairs Center, Dartmouth College, Hanover

Campbell S (2009) Improvement of clinical quality in English primary care before and after the introduction of a pay for performance scheme. PowerPoint presented at National Primary Care Research and Development Centre, University of Manchester, UK

Deci EL, Koestner R, Ryan RM (1999) A meta-analytic review of experiments examining the effects of extrinsic rewards on intrinsic motivation. Psychol Bull 125:627–668

Fast N, Berg N (1975) The Lincoln electric company. Harvard Business School Case 376-028

Leber H (2007) Chancengleichheit fördert ungleiche Einkommen. FAZ, 28 Dec 2007

Lenin W (1961) Werke Band 26, September 1917—Februar 1918. Dietz Verlag, Berlin

Lotter W (2008) Die neue Leistung. Brand Eins, pp 50–57

Meyer MW, Gupta V (1994) The performance paradox. Res Organ Behav 16:309–369

Nullmeier F (2000a) Politische Theorie des Sozialstaates. Campus Verlag, Frankfurt

Nullmeier F (2000b) "Mehr Wettbewerb!" Zur Marktkonstitution in der Hochschulpolitik. In: Czada R, Lütz S (eds) Die Politische Konstitution von Märkten, Opladen, pp 209–227

Rosa H (2006) Wettbewerb als Interaktionsmodus. Kulturelle und sozialstrukturelle Konsequenzen der Konkurrenzgesellschaft. Leviathan 34(1):82–104

Rosen S (1988) Promotions, elections and other contests. J Inst Theor Econ 144:73–90

Smith A (1759[1984]) The theory of moral sentiments. Liberty Fund, Glasgow

Smith P (1995) On the unintended consequences of publishing performance data in the public sector. Int J Publ Adm 18:277–310

Sprenger R (2002) Mythos motivation. Campus Verlag, Frankfurt

UBS (2005) Verwaltungsmanagement. Impulse zur Unternehmensführung, UBS Outlook

van Thiel S, Leeuw FL (2002) The performance paradox in the public sector. Publ Prod Manag Rev 25(3):267–281

Chapter 4
Happiness by Maximisation?

Kurt Bayertz

4.1 Two Distinctions

It is a popularly held belief that human action is always aimed at the realisation of a maximum. Let us call this the 'maximisation assumption'. It pertains to a descriptive assertion: an assertion about a factual tendency found in human beings and their actions. This assumption reared its head, for example, when the international financial crisis set in motion in August 2007 was attributed to human 'greed': to the desire for more and more and more, in particular more and more and more money. Whether or not this 'greed' is an ineliminable part of human nature or the product of specific social conditions, such as capitalism, is a matter of some controversy, however. Whereas in everyday use the maximisation theory is often cited with regret and critical intention, in large parts of the scientific community it is viewed as a neutral description of just the way human beings are. In some areas of psychology, biology and especially economics, it assumes a quasi-axiomatic status in explaining and predicting human action. *All* human action.

This maximisation assumption needs to be distinguished from the 'maximisation *principle*': the requirement that human action *ought to* aim at the realisation of a maximum. We occasionally come across this prescriptive assertion in an everyday context, but more frequently in the economic sciences and in decision theory, where it is deemed a principle of rationality. According to Gary Becker, for example, "everyone more or less agrees that rational behavior simply implies consistent maximisation of a well-ordered function, such as a utility or profit function" (Becker 1976, p. 153). Accordingly, 'acting rationally' means nothing other than: always choosing from the available options the one course of action which, when realised, will be linked to the greatest possible utility. Some philosophers have also appropriated this interpretation of rationality, as illustrated by the following passage from David Gauthier's book *Morals by Agreement*, in which

K. Bayertz (✉)
University of Münster, Münster, Germany
e-mail: bayertz@uni-muenster.de

H. Brockmann and J. Delhey (eds.), *Human Happiness and the Pursuit of Maximization*, Happiness Studies Book Series, DOI: 10.1007/978-94-007-6609-9_4,

he summarises the standard position of decision theory: "Practical rationality in the most general sense is identified with maximisation. Problems of rational choice are thus of a well-known mathematical type; one seeks to maximise some quantity subject to some constraint. The quantity to be maximised must be associated with preference; but the theory of rational choice defines a precise measure of preference, *utility*, and identifies rationality with the maximisation of utility. Utility is thus ascribed to states of affairs considered as objects of preference relations. The constraint under which utility is to be maximised is set by the possibilities of action. The rational actor maximises her utility in choosing from a finite set of actions, which take as possible outcomes the members of a finite set of states of affairs" (Gauthier 1986, p. 22). I shall return to some of the points mentioned here later on.

If we now examine this maximisation principle (MP) more closely and ask *whose* utility is to be maximised, a second distinction results. On the one hand (based on an objective concept of rationality), the principle can be interpreted as demanding that agents maximise the *universal* utility. This is a demand advocated by several moral philosophers, in particular the utilitarians. According to John Stuart Mill, the norm of utilitarianism "is not the agent's own greatest happiness, but the greatest amount of happiness altogether" (Mill 1969, p. 213). For utilitarians, MP possesses not only the status of a principle of rationality, but also that of a moral principle. Here the rational and the moral coincide. For each and every agent, MP amounts to a moral obligation to maximise the utility of *all* those affected by a course of action. For example, parents then have an obligation to ensure that their (future) children have a chance of enjoying the best life possible. "If couples (or single reproducers) have decided to have a child, and selection is possible, then they have a significant moral reason to select the child, of the possible children they could have, whose life can be expected, in the light of the relevant available information, to go best or at least not worse than any of the others" (Savulescu and Kahane 2009, p. 274). This explicitly includes application of the diagnostic and therapeutic options provided by gene technology in order to guarantee the best possible genetic make-up of (future) children.

On the other hand (based on a subjective concept of rationality) MP can also be interpreted as demanding that agents maximise their *own* utility. According to this view, it is not rationality and morality which coincide, but rationality and prudence; and MP is comprehended not as a moral principle, but as a prudential principle which governs actions or decisions. The hegemonic influence it currently possesses was acquired after it was adopted by the economic sciences in the 19th century and then formally developed in the book *Theory of Games and Economic Behavior* by John von Neumann and Oscar Morgenstern in the 20th century. In agreement with the economic tradition of the time, the authors took it as given "that the consumer desires to obtain a maximum of utility or satisfaction and the entrepreneur a maximum of profits". From this (empirical?) supposition they proceeded directly to the normative ascertainment: "The individual who attempts to obtain these respective maxima is said to act 'rationally'" (von Neumann and Morgenstern 1944, pp. 8–9). In the decades which followed, the principle was

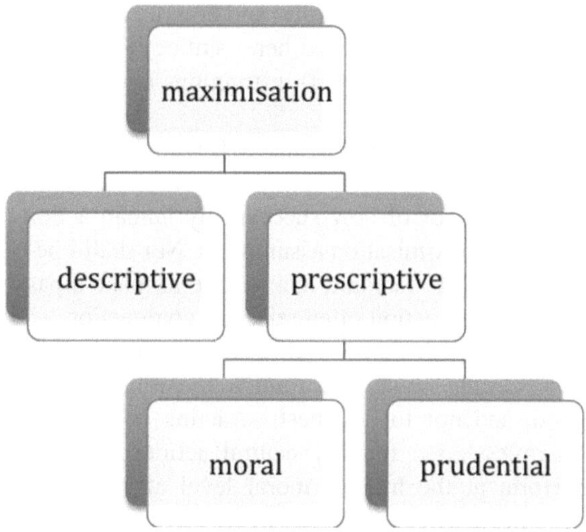

Fig. 1 Diagram of the two distinctions

increasingly viewed as a *general* principle of rational action, beyond the field of economics, and then codified within the framework of decision theory (Luce and Raiffa 1957). Although it soon became clear that it was based, in its formally elaborated, axiomatised form, on unrealistic prerequisites, it is still taken as a "gold standard" (Klein 2001, p. 104) for the rationality of decisions taken by all types of agents in all areas of life (Fig. 1).

4.2 Prudence and Happiness

We commonly say of an action that it is 'prudent' if it promotes the happiness of the agent. Viewed as a general principle of practical rationality, MP (in its prudential variant) states: a person wishing to be, to remain or to become happy must maximise. Or: the more we succeed in extracting the maximum from every situation, the happier we will be. This idea has been around for a very long time. It can be found in various Platonic dialogues, for example, where it is ascribed to the sophists. Let us take a look at the following plea: "The man who is to live rightly should let his appetites grow as large as possible and not restrain (kolazein) them, and when these are as large as possible, he must have the power to serve them, because of his bravery and wisdom, and to fill them with whatever he has an appetite for at any time… luxury, intemperance (akolasia), and freedom, if it is well supplied, this is virtue and happiness" (Plato 2009, pp. 491e–492d). Plato's dislike of the sophists is very apparent in this formulation, and he attempts to discredit them by linking them to MP. They are meant to appear as blind hedonists and inconsiderate egoists who strive for maximum happiness of their own, even if this harms others in the process. Whether or not the sophists really did advocate such a principle historically, and how we should interpret Plato's representation of them, cannot be explored here; suffice it to say that, for Plato, the principle represents (albeit misguided) instructions for happiness. And it is often still seen that way today.

I shall concentrate in the following on this prudential variant of MP. I shall therefore *not* be addressing the issue of whether or not human beings really are maximisers, or of how successfully human action can be explained or predicted using the maximisation assumption. Nor shall I be taking a closer look at the moral variant of MP. I should merely like to make a passing reference to the 'excessive demands' objection often raised in conjunction with utilitarianism. For utilitarians it is not enough simply to do good; rather, with each action one has to strive for the best possible result. According to the utilitarian view, a person who strives only for good, and not for the best, is acting not only irrationally, but also immorally. 'Harmless', i.e. morally neutral actions no longer exist; agents are obliged to perform at the highest moral level at all times. The extent of this obligation becomes particularly clear when we realise that it is the best possible result *for everyone* which is being demanded. Agents are thus required to demonstrate an impartiality which excludes not only a precedence for satisfying their own interests, but also excludes a precedence for privileging other persons close to them. Instead of financing the education of their own children, for example, parents with utilitarian motives would therefore have to finance the education of their neighbours' children if this would (probably) lead to greater overall utility. The objection that such demands are excessive has of course prompted a response from the utilitarians (or consequentialists), who have developed proposals for how such problems can be solved within their theoretical framework (Jamieson and Elliot 2009; Mulgan 2001). There is no room to discuss these proposals here.

What should have become clear by now is that the relationship between the moral and prudential variants of MP is a tense one. Although universal well-being and one's own well-being are not always irreconcilable, this is uncomfortably often the case. And when it is so, a decision about the ranking of the two variants has to be made: does universal happiness weigh more than the happiness of the agent? In the past, this tension has often constituted the main objection to the prudential variant of MP, the latter prescribing that immoral action be taken (at least occasionally) if this should prove necessary for our own personal happiness. Of course this is a legitimate objection; and yet its impact is limited. We could, after all, easily imagine a variant of MP which links maximisation to a condition, along the lines of: 'Maximise your utility, but only within the limitations of that which is morally permissible!' Such a variant would circumnavigate the objection of potential immorality, and we would be forced to conclude that, under this condition, maximisation was rationally required.

A far stronger objection to MP would emerge if it were possible to demonstrate that this principle is not only morally, but also *prudentially* counterproductive, in other words that maximising behaviour (independently of potential infringements of moral norms) is *not* conducive to the happiness of the agent. This objection also dates back to Plato, who attempted to justify it in detail in his *Republic*. Whereas

Plato put forward some very presumptive metaphysical arguments, the foundations for this objection in more recent empirical research are very different. Together with a number of colleagues, Barry Schwartz developed a catalogue of questions enabling test persons to be classified as "maximisers" or "satisficers". The behaviours and psychological conditions of the members of both groups were then investigated in several subsequent interviews and experiments. They revealed that "maximisers" have a significantly lower level of satisfaction, as well as a lower level of self-esteem, and that they are less happy and less optimistic. "Maximisers" regret their own decisions far more than the members of the other group and have a significantly higher tendency towards depression. In the study, extreme "maximisers" achieved almost clinical levels of depression (Schwartz et al. 2002; Schwartz 2004). It is worthy of note that "maximisers" achieve significantly higher incomes in their careers and yet are less content with their professional situations (Iyengar et al. 2006). Even though they achieve *objectively* better results through their decisional behaviour, their results are therefore *subjectively* worse.

It is not immediately apparent what this finding means. Firstly, it is of course possible that the empirical evidence is not correct. For example, authors of a later study were unable to confirm the negative results of the maximisers: "Our findings suggest that maximisers may not be so unhappy after all" (Diab et. al. 2008, p. 364). What is true, however, is that in this later study the maximisers also appeared not to be any happier than the non-maximisers, even though they supposedly should be. In addition, the later investigation also came to the conclusion that "maximisers" tend far more to regret their decisions later on than non-maximisers do; and no one could claim that regret is conducive to happiness. Other studies (Chowdhury et al. 2009, Parker et al. 2007) have confirmed the findings of Schwartz et al. (2002). Overall, it seems safe to conclude that these findings are not fundamentally wrong.

A second objection could be based on the fact that MP is a *principle*, whereas the findings of Schwartz et al. (2002) refer to a *decisional disposition* or *factual decisional style* (or more precisely: to persons who attribute themselves with having this decisional style). The principle dictates a prescriptive and universal rule which can be applied in practice more or less well. It establishes an ideal which is seldom achieved in real conditions. If a principle is applied suboptimally, then the principle cannot be held responsible for suboptimal results. This objection draws attention to an important difference: a principle is one thing; whereas a mental disposition, a decisional style, not to mention the actual decisions reached on the basis of the principle, are all quite another. And yet however much this difference may be justified conceptually, in the contexts of interest to us here its reach is limited. For (a) one will hardly be able to doubt the existence of correlations between the two sides of the difference. MP may be viewed as the 'rational reconstruction' of a maximising decisional style; or a factual decision may be viewed as a (more or less good) application of MP. And (b) the difference cited exists for *all* principles. It is probable that, in practice, principles of action are always applied suboptimally. Since this will also be true of the principles applied by the non-maximisers, the empirically ascertained relative differences between the two decisional styles cannot automatically be attributed to this factor.

With all due caution, these findings may thus be evaluated as indication that a decisional style aimed at maximisation is not conducive to happiness, and that in terms of happiness MP is thus counterproductive. That at least is the hypothesis upon which I shall be basing the following deliberations. But then what could this counterproductivity be ascribed to? Three possibilities require consideration.

1. *The concept of utility*. For reasons which cannot be gone into here, MP is based on a subjectivistic concept of value or utility, according to which 'valuable' or 'useful' are what the individuals in question *deem* to be valuable or useful; and it is based on an instrumentalistic concept of rationality, according to which 'rational' denotes the adequate choice of means to given ends, but with the ends themselves being beyond discussion (cf. Gauthier 1986, 25f). Apart from the fact that both prerequisites are philosophically substantive and contested, it could be that the reason behind the counterproductive effects of maximisation is that the wrong things are being maximised. Empirical findings exist which point in this direction (for an overview cf. Haybron 2008). Then the problem would no longer be maximisation itself, but the maximised goals.
2. *Summation*. MP prescribes maximisation in discrete situations. If we now perceive 'happiness' not merely as a state following on directly from a single decision, but as a temporally more or less prolonged state, ideally over an entire lifetime, then ensuring the rationality of a decision in a discrete situation is no longer sufficient. Far more, we have to view sequences of decisions, ultimately the entire sequence of all decisions made in one lifetime. The conventional approach states that lifetime happiness is compiled from the sum of discrete experiences of happiness (brought about by individual rational decisions). A person who maximises successfully in each discrete decisional situation will accordingly achieve the maximum happiness over his or her entire lifetime. Now this summation theory could conceivably be wrong; a decision-maker could maximise successfully in each individual situation and still not be happy or maximally happy. If this possibility cannot be ruled out, then MP cannot form the core of a comprehensive theory of practical rationality. Since MP does not provide a criterion for sequences of decisions, it contributes nothing to ensuring the sustainability of happiness.
3. *Maximisation itself*. Finally, the possibility also exists that there is something wrong with maximisation itself with regard to discrete decisions; that consistently maximising behaviour is in itself counterproductive in terms of happiness. I shall not pursue the first two possibilities any further in the following, concentrating instead exclusively on this third idea. I shall attempt to make plausible that MP (even in its less idealised interpretations) is problematic *as such*; or at least when viewed as a general principle of practical rationality.

4.3 Tendential Endlessness

If we question more closely what MP demands, we encounter a situation which is not particularly clear. In contrast to what might be expected in the light of the formal elaborateness of the literature on decision theory and economics, no uniform use of the term 'maximisation' or related terms such as 'optimisation' exists (cf. Klein 2001). According to a proposal repeatedly put forward by Amartya Sen, the maximisation postulate merely requires that, out of the options open to them, decision-makers choose none for which a better option is known to them by comparison (Sen 1997, pp. 746, 763; Sen 2000, pp. 483, 486). This proposal might be uncontroversial, but it is also not very helpful. For the idea that, out of a sum of given and known options, one should not choose one which is worse than another known and chooseable option follows on automatically from the conventional definition of the expression 'good', stating nothing other than that the thing thus denoted is 'preferable ceteris paribus'. A person who has understood what the expressions 'good' and 'better' mean therefore has to accept Sen's definition. This is (put kindly) analytically true or (less kindly) trivial.

In place of a definition, I shall start from an intuitively plausible example of the rationality of maximising *behaviour* and describe the difficulties which the maximiser encounters in pursuing it. I shall thus analyse the maximising decisional style and its consequences for happiness, and in so doing will view MP as an incitement to adopt this style in practice. Relativising the conceptual difference between a principle and its application in this manner seems to be legitimate in this context because a connection is to be established between the empirical findings mentioned briefly above and the 'logic' of MP. It should become plausible why application of MP is not conducive to happiness. Since in the relevant literature houses are often sold, I too shall discuss an example from the province of real estate. To this end, let us imagine that Mary would like to sell her house. She has had three offers: the first for €100,000, the second for €110,000 and the third for €120,000. Which offer should she accept? The obvious and intuitive answer would be that Mary should accept the third offer. It would be irrational to sell the house for €100,000 if she can get more. But *why* is that so?

There are several possible answers to this (rarely asked) question, but I would like to restrict myself to just one. It has two components. (1) The first component states: if something is good (has a 'utility'), then it would be unreasonable to wish to achieve less of it than is possible. Put another way: *if* g represents a good (utility, value, etc.) and *if* an agent can realise different quanta of g through different possible actions; *then* it is rationally required to choose the option which realises the largest possible quantum of g. This deliberation seems to give a strong justification for MP. It tells us why Mary ought to *maximise* the sale price; but not why she ought to maximise *the sale price*. (2) To answer this latter question, we have to examine Mary's preferences. *If* it is important to Mary to earn as much money by selling her house as possible, *then* she should accept the highest offer. Another scenario is also conceivable, however, in which the most important factor for Mary is her garden, her pride and joy, and that it will be taken care of properly once the house is no longer hers. It would then be rational to sell to the person who can provide her with the maximum guarantee of this happening; to sell to a passionate gardener. Maximisation takes place in both scenarios: in the one case with regard to price, in the other with regard to the commitment of the buyer to gardening.

Here we can see how the preferences of the agent determine the utility to be achieved; once that has occurred, it is then rationally required to maximise this utility. But we can also see how this simple example involves several prerequisites. *One* of these is that decision-makers can only maximise sensibly if they have a clear idea about their utility; and for this, in turn, they need to have a clear idea about their preferences. In standard decision theory, this prerequisite has been defined very precisely: accordingly, 'utility' results from a complete and transitive ordering of all preferences. In Mary's case, for example, it is assumed that as high a sale price as possible is more important to her than any other parameter (including whether or not the buyer enjoys gardening). Another prerequisite is that this order of preference does not change. Although it is obvious that one needs to have a sufficiently exact idea of what one wants in order to act successfully and maximise the success of one's actions, these prerequisites are patently excessive. Empirical evidence has confirmed that human beings possess such an order of preference only in exceptional cases, if at all. Under realistic conditions, therefore, a key prerequisite of the maximisation principle remains unfulfilled.

In the following I should like to examine another prerequisite in more detail. Mary has precisely three options (= offers on her house) and they are known to her. The intuitive plausibility of this example is obviously based on this prerequisite: *if* Mary more than anything wishes to achieve a high sale price; and *if* the three cited offers are available to her; *then* it is rationally required to choose the third one. And yet this conclusion is only compelling if the second prerequisite is fulfilled: in other words if exactly these three offers have been made to Mary and are known to her. Of course, we could also imagine a different number of offers with different sale prices; but in each case Mary can still only reach a maximising decision if a particular number of offers is *available* to her and *known* to her. In standard decision theory, this condition has repeatedly been underlined. We recall that Gauthier (1986, p. 22) formulated: "The rational actor maximises her utility in choosing from a finite set of actions, which take as possible outcomes the members of a finite set of states of affair". The decisional situation is thus presumed to be 'closed'.

Here the problems begin. The first becomes clear if we consider that usually when selling a house it is possible to wait, beyond the offers available at a certain moment, for other offers to come in, or to become active and seek other offers. One can, for example, put an announcement in the newspaper, advertise one's house on relevant websites or commission a real estate agent to find a further potential buyer who might be interested in offering €130,000 for the house. It should be clear that in very many situations, albeit not in all, the possibility of increasing one's options exists; and that this is probably even standard. Human beings are active creatures, not only in the sense of choosing actively from among given options, but also in the sense of seeking to change their options to their own advantage.

But if the options are no longer fixed and can instead be increased, Mary no longer has a reason to accept the highest of the three offers. And even if by waiting or actively seeking she found somebody willing to offer €130,000, the game would still not be over because, by waiting or seeking again, she might be able to find a buyer willing to offer €140,000 or €150,000, and so on... The problem is therefore that it would be irrational for Mary to accept *any* offer. In principle, the sale price of her house knows no limits: there *is no* highest price. If we view the problem from Mary's epistemic perspective, it becomes even clearer that in striving to achieve the maximum she can *never* come to a decision. Even if a maximum were to exist, she could never *know* that for sure. Therefore, even if nobody *will* ever de facto offer her more than €130,000, she cannot know this for sure and can therefore always hold out in the hope of achieving €140,000. Consequently: in contrast to the decisions of choice involved in 'closed' situations, maximisation becomes *fundamentally impossible* as soon as the possibility of increasing one's options comes into play.

Looked at more systematically, Mary is confronted with not one, but two decisions:

D_1 She has to choose one of the three cited options. The object of this decision is the prices offered by the three potential buyers

D_2 At the same time she has to decide between choosing one of these three options or waiting/seeking additional offers. The object of this decision is the number of options available.

Taking the two situations together, D_2 logically has to come first. Mary could not decide D_1 without eo ipso co-deciding D_2; this is not true in reverse. But when she (inevitably) makes her D_2 decision, a maximising Mary has to apply MP as a universally valid decisional principle. She is therefore rationally obliged to increase her options by waiting or seeking. Limitation to a fixed number of options is therefore not only unrealistic and artificial, but also directly contrary to MP. It therefore seems as if the rational decision-maker is forced to increase his options in a process which can never end; that he is sent on a path which cannot lead to a destination and is therefore endless.

Although this problem does not seem to occur in 'closed' decisional situations, a closer observation reveals that here, too, the decision-maker is led along a path which is endless. This brings us to a second problem. We conclude that the options of choice (whether there be a finite number or not) must be *known*. To the extent that this demand refers to the mere existence of options, it is trivial; of course one can only choose between options which one knows to exist. And yet the options also have to be 'known' in a more sophisticated sense: one has to be able to estimate the (expected) utility connected with choosing them. In Mary's case this is very easy: she knows that the utility increases with the increase in sale price and therefore has no problem in identifying the offer with the greatest utility for her.

But the situation is more difficult if Mary is primarily concerned not with the sale price, but with the future of her garden. She then has to find out how committed each potential buyer is to gardening. Since the relevant knowledge is useful to her, Mary will have to maximise it. It will hardly suffice to ask each buyer how passionate he or she is about gardening; instead she will have to research their horticultural interests and botanical experience with care. Even if she has only three potential buyers, this may involve a considerable amount of time and/or money: she will take a look at the buyers' present gardens or maybe engage a private detective to investigate for her. The more comprehensive, more precise and more reliable this information needs to be, the more effort will be required; it has the same tendency towards endlessness as the efforts to increase one's options.

If we now make ourselves aware of the fact that this problem of acquiring information occurs in nearly all realistic decisional situations (the exception being a decision between different prices for exactly the same good), then it becomes clear that in a practical and relevant respect the difference between 'closed' and 'open' decisional situations is more one of graduation than of principle. In 'closed' situations maximisers might not need to bother with increasing their options; but they are still rationally obliged to maximise the information available about each option.

These theoretical deliberations correspond to the behaviour of 'maximisers' as ascertained empirically. Several studies have found evidence of a widespread tendency in 'maximisers' to increase their options and to identify from the known options the 'best possible one'. Compared to non-maximisers, the consequences of this tendency are significantly greater decisional stress and a significantly reduced satisfaction with decisions reached (Chowdhury et al. 2009; Iyengar et al. 2006). Further confirmation of these findings would explain, at least in part, why maximising is not conducive to happiness.

4.4 External Constraints

Friends of maximisation will not be particularly impressed by this result and will attribute it to a misunderstanding. The consideration outlined above first implies the idea of an 'absolute' maximum and then predictably arrives at the insight that there can be no such thing. This is especially true of sale prices, of course. *Every* price can be numerically topped by another one; in absolute terms there can therefore be no highest price. And yet, within the context of the maximisation principle, it is not an absolute maximum which is meant. In standard decision theory, a finite number of given and known options is presupposed, from which the decision-maker then has to make his choice. It should now be clear why this supposition is necessary: from a fixed set of options there can be a maximum, which can then also be identified and chosen. In our example, Mary has precisely three options, of which one is the highest, and MP stipulates that this is the one to be chosen. MP is therefore aimed not at an absolute, but at a 'relative' maximum: relative to a set of given and known options.

We have seen, however, that such a limitation to given and known options is artificial; even that it is in contradiction to the 'spirit' of MP. An agent who is rational in the MP sense will be concerned with increasing the number of available options. And friends of maximisation will further argue that, in so doing, the decision-maker is *not* going off down an endless path. For we have not yet taken into account the fact that each instance of maximisation takes place under empirical conditions. Of course Mary can *hope* to find somebody through waiting or seeking who is prepared to offer €130,000 (or more) for her house. And yet, for all her hoping, she should not ignore actual market conditions. It could be the case that in the light of the actual housing market Mary's hope is phantasmal, that €120,000 already represents an extraordinarily good offer for her house and that she would therefore be well advised to take it and be content. Even though numerically there is an infinite number of prices higher than this sum, €120,000 could be the highest sum which will *really* be offered. Mary therefore has to reckon with this sum representing the real maximum.

The obvious counter-argument that the housing market can also change, that a higher price at a later moment in time cannot be ruled out, does not get us very far. For it is probable that Mary has not just *one* preference (a high sale price), but others besides. This has already been made clear in the abovementioned call for an order of preference, in which the various preferences are viewed systematically. In particular, all human decision-making and actions take place within a limited time. Life is finite and Mary will not want or be able to wait until she achieves the maximum price sub specie aeternitatis; instead she will accept the highest offer made to her within a period of time she herself will determine. Generally speaking: even if it is right at the top of an order of preference, maximisation of a single preference will always be limited by other preferences.

In the real world, every instance of maximisation occurs under certain constraints, of which we have addressed two types: (1) external factual conditions imposed (economically speaking) by the market and (2) internal factors which arise from the multitude of preferences held by all agents. Additional constraints can also be cited which do not require further discussion here: moral norms, for example. Even if MP sets decision-makers abstractly onto a path of endless progression, in actual fact they are subject to various types of constraint which will counteract the endlessness of this progression. Under real life conditions, maxima do not extend sky-high.

The technical expression for maximising under limited conditions is 'optimising'. Mary, for example, has to determine a period of time within which she will decide to take the maximum offer available to her by then. If we now assume that this period of time can be chosen freely (within certain limits), then by lengthening or shortening it she can influence the probable level of the achievable sale price. The optimisation process therefore has (at least) two adjustable screws: Mary can try for a higher price but maybe have to wait longer; or she can sell fast but maybe not get such a good price. Striving for a maximum sale price is thus countered by a time preference. At a 'technical' level the problem of endlessness has thus disappeared. MP is no longer directing decision-makers down a path which will never end.

But what impact does application of MP have from the perspective of the agents or decision-makers? Consistent efforts to apply this principle will place them in an ambivalent situation. On the one hand, the principle makes high demands on them. It requires the greatest possible clarity with regard to their own preferences; a perfect overview of all the available options, including the utility (presumably) linked to each one; as well as an error-free calculation of that utility. It may be possible to characterise this as an ideal which—although unattainable—indicates the direction of searching and striving. Then the function of MP would consist in a reminder not to slack in one's efforts to reach the best decisions possible and always at least to strive for the maximum. The reward which MP promises for this effort is correspondingly high. Agents following this principle are guaranteed that they will always achieve the best possible result and lead the best possible life. ('Best possible' not in an absolute sense, but in the relative sense explained above.) The expectations harboured by decision-makers on appropriating MP and attempting to apply it are no doubt correspondingly high.

In contrast to this is the experience of decision-makers in attempting to apply MP. Generally speaking, they are shown a path along which the only stopping point is one where progress is terminated. MP only foresees a clear stopping point in cases such as Mary's where there is a finite number of options, of which one is the highest. These cases are seldom, however, and not typical. With regard to the multitude of realistic decisional situations with which we are confronted, MP shows us a path which does not have an *internally* defined end. The point at which we stop all further searching and come to a decision is marked out by external and contingent constraints. Either it is the (in the most general sense of the word) market conditions which lead Mary to sell her house at a particular price; or it is other preferences of her own which guide her. In this second case we are also concerned with external contingent factors to some extent: for although the limiting factors are also her own preferences, they are not ones she wishes to maximise. In relation to her desire to achieve as high a price as possible, her limited time budget is an external and contingent constraint.

The deciding and acting subject therefore *always* has to be content with a compromise which is enforced through the finite nature of life, through the limitation of resources or through pressure from another preference. Taking MP as a basis, the subject can therefore never reach a decision which, in strong terms, is *his or her own*; and with which he or she can therefore be wholeheartedly satisfied. This is not just a theoretical presumption, but is empirically reinforced by the findings outlined briefly above. Amongst other things, consistent efforts towards maximisation promote decisional stress and regret and are therefore counterproductive in terms of happiness.

4.5 Summary and Conclusion

1. The deliberations in this chapter do not rule out the possibility that it may be rational in certain situations and under certain conditions to maximise. Mary's example could describe just such a situation. But situations of this type are the exception. A false assumption that they represent the rule is one of the reasons why MP is still persistently deemed to be a general principle of practical rationality.
2. In its formally elaborated, axiomatically polished form, MP is not universally applicable because it entails unrealistic preconditions, both with regard to the deciding subject and to the decisional situation. This has largely been recognised. But it is not sufficient to lower one's sights from the ideal simply in the interests of practicability. Even in its less idealised versions, MP prescribes a decisional behaviour which, on the one hand, promises decision-makers maximum results and yet, on the other, makes each decision reached seem like a compromise which has been imposed by contingent constraints.
3. These deliberations permit us to surmise that a consistently maximising decisional style is (a) incompatible with the personal autonomy of the agent, at least in a demanding sense of 'autonomy', and (b) is not actually conducive to the happiness of the agent. Empirical evidence exists to support this supposition.
4. If we assume the theory to be correct that a decisional style is only prudentially rational if it sustainably promotes the happiness of the agent, then the deliberations laid out here lead us to conclude that consistent maximisation is not rational.

Acknowledgments I would like to thank the "Hanse Wissenschaftskolleg" in Delmenhorst for providing me with a fellowship in 2009/10 on which I began the work on the problems of maximisation. I continued this research as part of my ongoing project at the "Center for Advanced Study in Bioethics" at the University of Münster. I am grateful to my colleagues at this centre as well as Nikola Kompa (Osnabrück) and Aloys Prinz (Münster), for helpful comments on earlier versions of the present paper. The German text was translated into English by Sarah L. Kirkby.

References

Becker GS (1976) The economic approach to human behavior. University of Chicago Press, Chicago

Chowdhurdy TG, Ratneshwar S, Mohanty P (2009) The time-hurried shopper: exploring the differences between maximizers and satisficers. Mark Lett 20:155–167

Diab DL, Gillespie MA, Highhouse S (2008) Are maximizers really unhappy? The measurement of maximizing tendency. Judgment Decis Making 3:364–370

Gauthier D (1986) Morals by agreement. Clarendon Press, Oxford

Haybron DM (2008) The pursuit of unhappiness. The elusive psychology of well-being. Oxford University Press, Oxford

Iyengar SS, Wells RE, Schwartz B (2006) Doing better but feeling worse. Looking for the "best" job undermines satisfaction. Psychol Sci 17:143–150

Jamieson D, Elliot R (2009) Progressive consequentialism. Philosophical Perspectives 23:241–251

Klein G (2001) The fiction of optimization. In: Gigerenzer G, Selten R (eds) *Bounded rationality. The adaptive toolbox.* MIT Press, Cambridge, pp. 103–121

Luce RD, Raiffa H (1957) Games and decisions. Introduction and critical survey. Wiley, New York

Mill JS (1969) Utilitarianism. In: Robson JM (ed) Essays on ethics, religion and society. Volume 10 of collected works of John Stuart Mill. University of Toronto Press, Toronto

Mulgan T (2001) The demands of consequentialism. Clarendon Press, Oxford

Parker AM, de Bruin WB, Fischhoff B (2007) Maximizers versus satisficers: decision-making styles, competence, and outcomes. Judgement Decis Making 2:342–350

Plato. *Gorgias*. translated with notes by Terence Irwin (2009) Clarendon Press, Oxford

Savulescu J, Kahane G (2009) The moral obligation to create children with the best chance of the best life. Bioethics 23:274–290

Schwartz B (2004) The paradox of choice: why more is less. Ecco, New York

Schwartz B, Ward A, Lyubomirsky S, Monterosso J, White K, Lehman DR (2002) Maximizing versus satisficing: happiness is a matter of choice. J Pers Soc Psychol 83:1178–1197

Sen A (1997) Maximization and the act of choice. Econometrics 65:745–779

Sen A (2000) Consequential evaluation and practical reason. J Philos 97:477–502

von Neumann J, Morgenstern O (1944) Theory of games and economic behavior. Princeton University Press, Princeton

Chapter 5
Maximization and the Good

Valerie Tiberius

> By the principle of utility is meant that principle which approves or disapproves of every action whatsoever, according to the tendency it appears to have to augment or diminish the happiness of the party whose interest is in question: or, what is the same thing in other words, to promote or to oppose that happiness. I say of every action whatsoever, and therefore not only of every action of a private individual, but of every measure of government (Bentham 1996).

5.1 Introduction

The Classical Utilitarians, Jeremy Bentham and John Stuart Mill, famously thought we should maximize happiness. But their claim that we ought to maximize happiness can be misunderstood today unless we take care to see why they thought happiness ought to be maximized. Classical Utilitarianism is the combination of two theories; it combines a hedonistic theory of the good with a consequentialist theory about right action. Hedonism is the view that the one and only good is happiness defined as pleasure and the absence of pain. Consequentialism is the view that the right thing to do is whatever produces the most good. Taken together these theories tell us that the right action is the one that maximizes happiness.

Bentham and Mill spend a good deal of time considering what happiness is. They also spend time defending the idea that happiness is the good. But they spend very little time indeed explaining why we should maximize happiness and almost none explaining why we should maximize the good. The idea that we should maximize the good is taken to be obvious. After all, if you know what is good—that is, what is objectively worth pursuing—naturally you should produce as much of it as you can. What possible reason could there be for you to stop at, say, fifty percent of the good available when you could produce more of it? One may have doubts about the idea that there is something we know to be objectively worth pursuing. But on the assumption that there is such a thing and that we know what it

V. Tiberius (✉)
Department of Philosophy, University of Minnesota, South Minneapolis, MN, USA
e-mail: tiberius@umn.edu

H. Brockmann and J. Delhey (eds.), *Human Happiness and the Pursuit of Maximization*, Happiness Studies Book Series, DOI: 10.1007/978-94-007-6609-9_5,

is (assumptions Bentham and Mill thought were true), then the idea that it should be maximized follows quickly.

The important thing to notice here is that it is not obvious that happiness should be maximized. What's obvious is that the good should be maximized. Bentham and Mill thought the idea that the good *is* happiness was one that had to be argued for. Notice also that there are two features "the good" must have for it to be the kind of thing about which it makes sense to say that it should obviously be maximized. First, it must be all-encompassing. *The* good is everything that is good; it's not *a* good, after all. There are no good things that are not part of *the* good. If the good in our theory were not all encompassing—if, say, there existed several goods and our theory only covered one of them—it wouldn't make sense to maximize, because we might produce more of one good at the expense of the others. Second, the good must be commensurable, that is, there must be a single standard by which we can measure more or less of it. If this were not the case, the good could not be maximized because there would be no fact of the matter about whether we have chosen the state of affairs that maximizes the good if we cannot make meaningful comparisons between different states of affairs in terms of how much good they contain. Now, happiness has both of these features (or so the Utilitarians supposed). Bentham and Mill thought that happiness was the only good and that anything else that seemed good was good as a means to happiness or as a constitutive part of it.[1] They also thought that happiness could be quantified and amounts of it could be compared.[2]

So, the only thing it makes sense to maximize is the all-encompassing, commensurable good. According to the Classical Utilitarians, pleasure fit the bill, hence they were maximizers. But is hedonism the right view about the good? Not many think so. Is there a better view about the nature of happiness or well-being that might underwrite the maximization principle? In the next section of this chapter, I will discuss some standard views about happiness, or what is now more commonly called "well-being", in philosophy and psychology and consider whether these views posit something that is plausibly thought of as *the good.* According to these theories, maximization is not the simple matter Bentham and Mill took it to be. Next, in section three, I discuss a deep problem with maximization that comes from an insight that the Ancients had about the good life for a human being. Finally, in the last section I'll consider the implications of a plausible pluralistic theory of well-being, one which recognizes the wisdom of the Ancients, for the maximization principle. The theory I defend is the Value Fulfillment Theory (VFT), which attributes intrinsic prudential value to a number of different items in virtue of their being valued by the well-being subject. Pluralistic theories such as this one make maximizing problematic because they do not

[1] Mill 1979, in particular, argues extensively that virtue is good as a means or a constitutive part of happiness.

[2] Mill also thought that pleasures differed in degrees of quality (Mill 1979). This is an important difference between Bentham and Mill, but it need not concern us here.

provide a single target and, further, because they include items (such as friendship) that are not appropriately maximized.

5.2 Well-Being and the Good

Before we get to talking about what happiness is, a point about terminology is in order. Many philosophers these days make a distinction between happiness and well-being according to which the former is a psychological state and the latter includes everything that is prudentially good for a person, which is not necessarily limited to psychological happiness (Haybron 2008). To elaborate, "happiness" is understood as it is in the social sciences as some kind of positive psychological state such as pleasure or positive affect. "Well-being", on the other hand, is understood to encompass everything that could make a life go well for a person. Well-being (but not happiness), then, might be defined in terms of certain objective goods or the perfection of human capacities. Of course, it could be that the best theory of well-being is one that takes it to be identical with happiness. The point is that there is a *conceptual* difference between happiness (a life that feels good) and well-being (a life that is well lived). For the sake of being in line with what I take to be an emerging consensus on this view in philosophy,[3] I am going to use the word "happiness" to refer to positive psychological experience and "well-being" to describe what is gained by a person who lives a good life for her in the most general sense.[4] The distinction is useful, because it allows us to think clearly about the relationship between a life that goes well for a person and positive psychological states. Though it may be the background assumption in the social sciences, the idea that the best life for a human being to live is one that feels good is one view among many; it is certainly possible to think otherwise as did most of the Ancients (Annas 1995).

So, turning to the available theories of well-being, let's start with hedonism. The view that a prudentially good life is one in which a person has the most pleasure and the least pain possible has the advantage that it makes well-being something that can be measured. Kahneman claims to measure what he calls "objective happiness" and what he means by happiness is pleasant states of consciousness (Kahneman et al. 2004). But when it comes to being all-encompassing, hedonism comes up against some serious obstacles. It just seems fairly

---

[3] There is not complete consensus, however. Julia Annas for example, argues that the word "happiness" should be used to describe what is achieved in the best human life (Annas 1995).

[4] The point of talking about a good life *for* a person, or a *prudentially* good life is to distinguish the value of well-being from moral value. It's at least conceptually possible for a person to live a morally good life that isn't good for her (for example, the life of a person who sacrifices her life for a moral cause), or (somewhat more controversially) for a person to live a life that is good for her but not morally good (for example, Gaugin who abandoned his family to paint Tahitian natives).

obvious that there are other things that are important to how a person's life goes besides pleasure and the absence of pain. The philosopher Robert Nozick (1974) illustrated this point with his experience machine thought experiment. Nozick has us imagine being given the option of hooking up to a very advanced virtual reality machine that would guarantee us a life that contains more pleasure than the life we would live should we decide not to hook up. If pleasure were the only thing that was important to us, we should not hesitate to hook up to the machine.[5] But many of us do hesitate and Nozick thinks the explanation is that there are other things we value besides pleasure, for example, being in touch with reality or *doing* certain things rather than just thinking that we are doing them.

In philosophy, a popular turn after rejecting hedonism was toward preference satisfaction theories (Brandt 1979; Griffin 1986; Railton 1986). According to such theories, well-being is getting what you want. Put like this, it's easy to see what's wrong with the view. We often want things that turn out to be bad for us, and we often don't want things that turn out to be good for us. To take a very simple sort of example, my sister has always had a great desire for delicious, French baguette, but it turns out that she has a serious gluten allergy. Eating a baguette satisfied her desire but did not contribute to her well-being. Most philosophers who favor preference satisfaction views now defend *informed* preference satisfaction theories, according to which what is good for you is getting what you would want if you had the relevant information.[6]

Is happiness as the satisfaction of idealized preferences or desires a good candidate for maximization? Not obviously. Idealization introduces serious obstacles to empirical investigation. It is not so difficult to ascertain what people want. We can ask them, or we can use a behavioral proxy for desire such as willingness-to-pay in order to ascertain what people want and how much. But how do we investigate what people *would* want if they were fully informed? Moreover, even if we could make reasonable inferences about what people would want from what they actually want, what would we actually maximize? According to this theory, the good is not the *feeling* of satisfaction; it is, rather, satisfaction in the sense of the achievement of the wanted object. So, a person who has informed preferences for fine wine, good health, and high quality National Parks is doing well when she has fine wine, good health and high quality National Parks. But what if the preference the person would have if fully informed are not the same as

[5] We are supposed to put aside practical worries about the machine failing or the intentions of the "super-duper neuroscientists" who run the machine. We are also supposed to imagine that others have the same option so that our decision to hook up or not will not cause others to experience less pleasure. By defining the case as he does, Nozick is trying to isolate our intuitions in order to focus our attention on the question about whether pleasure is the only thing that matters to us.

[6] But not all; there are some who defend actual preference theories (Heathwood 2006; Keller 2004). Notice that actual preference theories can make use of a distinction between what is wanted intrinsically and what is wanted as a means to something else in order to deal with at least some apparent counter-examples.

her actual preferences? Should we provide high quality National Parks even if nobody *actually* wants them? More difficulties arise when we think about how informed preferences would be commensurated. How do we compare these different preferences? We could do so according to the strength of the preferences, but then we have an even more daunting task, which is to figure out not just what someone would want if they were fully informed but exactly how much.

Perhaps all these practical problems could be solved. We would still have to ask whether informed preference theories are all-encompassing. Do these theories include all the goods that there are? A number of philosophers have argued that these theories do not actually line up very well with what is good. This is because, they argue, the process of becoming informed (understood as the process of reflecting vividly on the facts) is not guaranteed to track what is good for a person (Rosati 1995a; Velleman 1988). It could be that what would really contribute to your well-being is not something you would want if you were fully informed, but some surprising experience that changes the preferences that you have or would have even if informed. For example, if you are an anxious and cautious person you might prefer, even if fully informed, not to risk trying a new dance class because the fear of embarrassment persists even after reflection on the facts. In this case, it might be better for you to take a drug that would reduce your anxiety and enable you to experience the pleasure of dancing than it would be to act on your informed preference; informed preference theories seem unable to make sense of this possibility.

Life satisfaction theories of well-being have similar kinds of problems. Life satisfaction theories can be understood in two ways. Psychologists think of well-being as defined in terms of actual life satisfaction, which is a positive overall subjective assessment of how your life is going for you. The view that life satisfaction is a main component of subjective well-being is very popular in psychology and has formed the basis of a large body of empirical research.[7] Actual life satisfaction can be measured, but does the subjective feeling of life satisfaction really encompass everything there is to a good life? The view of psychologists is illuminating here. Ed Diener, the father of life satisfaction theory in psychology, does not even think that life satisfaction is sufficient for *subjective* well-being, let alone for the even broader notion of well-being we are concerned with here (Diener et al. 2003). Insofar as we are concerned about the good life for a person (as opposed to the subjectively good life), we may want to modify the view that well-being consists in life satisfaction *tout court*, which is, in fact, what philosophers who defend life satisfaction theory argue that we should do.

According to the main proponent of this theory in philosophy, Sumner (1996), life satisfaction is an overall positive assessment of one's conditions of life and it constitutes well-being when it is "authentic". Life-satisfaction is authentic when it is informed and autonomous. These conditions guarantee that the satisfaction in question truly represents the agent's own subjective assessment, thereby ensuring

[7] For an informative recent anthology see Eid and Larsen (2008).

that the right things are encompassed by her well-being. Importantly, for Sumner, these conditions also solve what he thinks is the biggest problem for subjective theories, the problem of adaptive preferences. Briefly, the problem is that we can be satisfied with very little if we have adapted to oppressive circumstances. A person who has been raised to believe she does not deserve much from life may be satisfied with her impoverished state, but, Sumner thinks, this does not make her impoverishment good for her. On his view, if the person would not be satisfied with her life were she to know why it is that she has so little (or how little she has in comparison with others), or were her attitudes not shaped by an oppressive society, then she has not actually achieved well-being. The problem of adaptive preferences can be seen as a challenge to the scope of subjective theories of well-being, or, in other words, to whether they encompass all and only those things that are good for a person. The worry is that subjective responses that have adapted to oppression do not deserve to be thought of as part of a person's good. A theory like Sumner's solves this problem by adding constraints on which subjective responses count as well-being.

Sumner thinks that well-being, as he defines it, is an all-encompassing value, but because his view involves idealizing conditions it runs into some of the same problems with respect to measurement as informed preference theories. Authentic life satisfaction may be commensurable in principle, but it will be very difficult to make assessments of *authentic* life satisfaction in practice. Of course, this is a practical problem that could be solved by better measurement techniques and more careful operationalizations of well-being. Perhaps we could find a way to measure and commensurate idealized subjective psychological states such as informed preferences and authentic satisfactions. This would clear the road for maximization, but we would quickly confront another obstacle.

5.3 A Deeper Problem: The Wisdom of the Ancients

A natural response to the thought that the good must be all-encompassing is to embrace pluralism: if it's difficult to find one thing that captures everything about the good, why not say that the good is a composite? This is just what a pluralist theory of well-being tells us: the good life for a person is made up of a number of different values. According to Aristotle, for example, the happy person has all the virtues and acts on them, and also has friendship, sufficient material resources, and some luck. Psychologists working on well-being also seem to embrace forms of pluralism. Ed Diener, as mentioned above, thinks that subjective well-being consists in global life satisfaction, domain satisfaction, positive affect, and low negative affect (Kesebir and Diener 2008). Seligman's (2002) theory of authentic happiness includes positive emotion, engagement, and meaning. And Carol Ryff and Burton Singer (Ryff and Singer 1998) take psychological well-being to include autonomy, personal growth, self-acceptance, life purpose, mastery, and positive relatedness.

Pluralism causes obvious problems for commensurability. As soon as a second good enters the picture, we no longer have a clear target for maximization. Now there are two things to be maximized and questions will arise about what happens when you can produce more of one good by sacrificing another. Pluralism also causes a less obvious problem, at least when we take seriously some of the ideas the Ancients had about the ingredients of a good human life. From the point of view of the Ancients, not everything worth having in a life is something that can be maximized. Julia Annas puts it this way:

> If my ethical aim is to produce a good, or the best, state of affairs, then it is only rational to produce as much as possible of it. But ancient ethics does not aim at the production of good states of affairs, and so is not tempted to think that rationality should take the form of maximizing them. Rather, what I aim at is my living in a certain way, my making the best use of goods, and acting in some ways rather than others. None of these things can sensibly be maximized by the agent. Why would I want to maximize my acting courageously, for example? I aim at acting courageously when it is required, I have no need, normally, to produce as many dangerous situations as possible, in order to act bravely in them (Annas 1995).

We can distinguish two points here. First, some valuable things are such that to treat them appropriately is not to maximize them. Second, living a good life requires figuring out how to treat different values appropriately and fit them together into your life: it requires balancing, rather than maximizing.

Friendship serves to illustrate the first point. Friendship is undoubtedly part of happiness, but it does not make sense to maximize friendship. Particular friendships are to be cherished and nourished, not maximized. And friendship in general is not to be maximized either. It's not even clear what it would mean to maximize friendship: would it be to have as many friends as possible? Or to make the friendships we have as intense as possible? Neither of these strategies is obviously the best strategy for living a good life. Annas's example of a value that it does not make sense to maximize is courage. Virtue in general isn't sensibly maximized. Developing our character is about responding appropriately to complex circumstances as they arise; it does not require acting on a particular virtue as much as possible, nor does developing a single virtue mean using it all the time or single-mindedly. Virtues work together, after all; compassion tempers justice, justice informs courage, and so on. Even if we could make sense of maximizing our own virtuous character traits, it surely does not make sense to maximize other people's virtue. If having a good character is part of what happiness is, we can't aim to produce the greatest happiness for the greatest number.[8]

The second point could be put this way: the development of agential capacities is required for living a good life, particularly if a happy life is understood in pluralist terms, and agential capacities (like virtues) are not an appropriate target

---

[8] Mill, of course, would disagree, as I mentioned above in fn. 1. But Mill's view that virtue is a part of happiness makes his hedonism into something quite different from Bentham's and I would argue that it is ultimately difficult to see how Mill's rather sophisticated conception of human happiness is something that can be maximized in the way that Bentham's pleasures could be.

for maximization.[9] If a happy life contains many incommensurable (or at least difficult to commensurate) goods, then living a happy life requires the ability to compare values and decide when it is worth sacrificing one for the sake of another. We need to be able to decide what to value, what goals to pursue, which preferences to indulge and which to squelch. For this, we need to be able to make value judgments and choices about what is worthwhile and what isn't. In the likely case that we find more than one thing to be worthwhile, we also need to be able to decide what matters more or less, to prioritize, and to know what to sacrifice when necessary.

One of the main capacities required here is the capacity to specify what we care about in a way that turns a vague value such as "accomplishment" or "relationships" into a determinate goal (Richardson 1997). After all, we don't value relationships and accomplishment in general, we value particular relationships and success in particular projects. The specification of ends helps to resolve conflicts as we seek to specify general values in ways that it is possible to pursue more than one at a time. Consider the familiar conflict between the values of career and family. For most people, to live a happy life, we must figure out a way of understanding these values such that we can appreciate the rewards of both. This may require altering our standards for what counts as having a successful career, or being a good parent, son or daughter, or it may require giving up other values we have that we don't take to be as important.

The wisdom of the Ancients is that a good life includes many things and that *living* a good life is an activity we engage in to try to figure out just what these good things are, put them together, and give each of them their due.[10]

5.4 Maximization in its Place

There are things we can measure and compare, but these things are not plausible candidates for the all-encompassing good that is worthy of maximization. As we move toward a theory that does encompass all the important facets of the human good, we find ourselves with a notion of the good that it doesn't really make sense to maximize. Should we just forget about maximizing altogether? I think we should proceed cautiously here. Even if well-being, according to the best understanding of it, is not the kind of thing we can maximize, this does not mean that more isn't often better. It is. But when and where and in what way more is better needs to be informed by a sophisticated understanding of the human good. In this

[9] The idea that agential capacities and what I call the virtues of "reflective wisdom" are necessary for living a good life is an idea I develop in more detail elsewhere (Tiberius 2008).

[10] Subjective theories of well-being like the informed desire theory or life satisfaction theory may be able to accommodate these insights by arguing that the plurality of values identified by the Ancients are reliable *causes* of well-being. But if they do this, given the nature of the causes, they will not end up with a simple story to tell about maximization.

final section I'll suggest a way to understand well-being that will help us go forward.

The theory I favor is the Value Fulfillment Theory of well-being, which says that to live well is to succeed by the standards of your own appropriate values. The Value Fulfillment Theory has the same structure as an informed preference theory, but it replaces preferences with values and the standard of full information with standards of appropriateness.[11] I'll expand on these differences in what follows, but in short, according to VFT, if your appropriate values include your own enjoyment, relationships with family and friends, accomplishing something in your career, and contributing to certain morally worthwhile projects, then your life goes well for you insofar as you have good relationships and career success, make a moral contribution and enjoy what you're doing.

Values (in the sense relevant to well-being) are the ends we take to give us reasons to act in the context of deliberating about and assessing how our lives are going. We endorse or avow our values as things that it makes sense to care about, plan for, pursue, or promote.[12] I mean to be very inclusive about what counts; values can include activities, relationships, goals, aims, ideals, principles, and so on. There are several distinctive features of the values we take to give us reasons. The first thing to notice is that to value something is to care about it in a particular way, and to care about something is, at least in part, to have some positive affective orientation toward it.[13] Other things equal, we are *motivated* to pursue or promote the values to which we are committed and we are disposed to react emotionally when these values are helped or threatened. For example, if I value being teacher I will be motivated to help my students learn, to feel proud when I receive good teaching evaluations and disappointed when my students sleep during class. Typically, valuing involves patterns of disposition to act and to feel. So, values are the central objects of these coordinated patterns.[14]

But there is more to a value commitment than motivation. If value commitments were simple motivational or affective states, they could not have the functions that they seem to have in our lives. If our value commitments are going to serve as the basis for deliberation and planning and for assessments of how well

---

[11] A pre-cursor to this theory is developed in Tiberius and Plakias (2010). I defend the theory in detail in my "Well-being for Uninformed: Prudential Reasons and the Value Fulfillment Theory". It is listed at the end as "unpublished ms."

[12] For a more thorough discussion see Tiberius (2008); in that work I talk about taking our values to be "justified" rather than "appropriate" as I now prefer. Values as I intend them are similar to what psychologists would call "life goals": "specific motivational objectives by which a person directs his life over time" Schmuck and Sheldon (2001, p. 5).

[13] For an illuminating discussion of caring see Jaworska (2007). I agree with Jaworska that caring does not have to involve reflectiveness (one of the main theses she defends in this article), but I do not take caring and valuing to be quite the same thing, as will become clear.

[14] If we take 'desire' in the broadest sense, where a desire is any pro-attitude, then valuing is a species of desiring and the value fulfillment theory of well-being is a species of a desire theory. It's not clear that this is a helpful way of thinking about 'desire', though it is true that VFT shares some of the advantages of desire theories in virtue of its similarity to them.

our lives are going, they must include more than good feelings. Not every pro-attitude plays an important role in planning and in the assessment of our lives. Some of our motives are ones we wish we did not have, and would be better off without; for example, the addict's desire for heroin, or the fleeting urge to jump from a tall building when one realizes that one could, do not seem like desires that provide prudential reasons. What we are motivated to pursue does not automatically give us reasons, then, even from our own point of view.[15] True value commitments have two other features that allow them to play the role that they do: stability and justification.

If our values were not relatively stable they would not help us in constructing plans to achieve the many rewards that can only come from sustained commitment. But too much stability, or stability for the wrong reasons, isn't desirable either. Our values must also be held appropriately if they are to be normative for us. Values are subject to *standards* of appropriateness. In the context of a subjective theory of well-being, appropriateness is going to be a matter of how one's values fit together as opposed to whether or not they meet some objective standard. We can wonder, for instance, if our values are informed, or if they are emotionally suited to us. We can also wonder if they reflect an ideal of the person that we find compelling (Rosati 1995b). The fact that we take our values to be subject to norms of appropriateness helps them to play the special role that they play in our lives. A person who values her relationship with her partner (on my characterization of what a value is) takes this value to be appropriate and this sustains her thinking that this relationship gives her good reasons to do things such as remembering her partner's birthday.

So, values are reason-giving, motivating, and relatively stable. These three features parallel three important features of well-being, which is also reason-giving, connected to the subject's motivations, and relatively stable. This is an important advantage of the Value Fulfillment Theory. Values are also especially related to subjects, unlike mere desires. People identify themselves in terms of their values and values are, by definition, of particular importance to people from their own point of view. Values have special relevance to well-being, then, insofar as well-being is a value that has a special relationship to the subject. Another benefit of the Value Fulfillment Theory, for our purposes, is that it provides a theoretical rationale for taking seriously the diverse range of psychological studies on well-being. According to the Value Fulfillment Theory, life-satisfaction and "objective happiness" are both components of well-being because people appropriately value them. If people also value some of the things that the Ancients thought were components of a good life (e.g., friendship and virtue), then those things too are components of well-being. VFT gives us a framework for

---

[15] I want to avoid taking a stand on whether desires necessarily give us *some* reason to act, however trivial. For my purposes, it is enough to say that desires simpliciter do not track prudential or well-being related reasons, reasons which are due some consideration in deliberating about what to do. Throughout the paper, then, when I talk about reasons I mean to refer to such prudential reasons.

understanding what matters and why. It allows for pluralism about the components of a happy human life, but it provides a unifying explanation for why certain components make the list.

Of course, Sumner's authentic life satisfaction theory also provides a rationale for taking into account a plurality of components of the good life, because these components are *causes* of life satisfaction. The difference here is that the authentic life satisfaction theory makes friendship, parenthood, or artistic expression (for example) good for a person instrumentally: these things are good in virtue of their causing life satisfaction. According to VFT, however, friendship, parenthood, artistic expression and the experience of life satisfaction itself are good for us intrinsically so long as we value them. If we also value positive affect and the experience of life satisfaction, which most of us do, then they are good for us too. For the purposes of deciding what well-being measures to use, this difference might not matter very much. The difference between the two theories matters when we confront conflicts between values. Life satisfaction theory implies that the way to resolve conflict is to opt for whatever produces more life satisfaction. That is the point of defining well-being in terms of life satisfaction. According to VFT, however, we should resolve conflicts by thinking about which values are more important to us in the context. In some cases we might take pleasure (positive affect or low stress) to be more important than life satisfaction and this is not a possibility that the life satisfaction theory can make sense of.[16]

I think the Value Fulfillment Theory is a worthy contender as a theory of well-being, but this is not the place for a full defense of it.[17] Those who favor other theories can accept the implications of VFT for maximization as long as they share two basic assumptions: (1) pluralism about the list of ingredients of a happy human life, and (2) the view that some of these ingredients are not appropriately maximized. I take it that many eudaimonists, preference satisfactionists and life satisfactionists could accept these assumptions, though they will have different theoretical explanations for them.

We can now turn to consider the guidance that the Value Fulfillment Theory provides about maximization. First, according to VFT, the maximization of one good should always be constrained by other goods that contribute to happiness. As long as pleasure is not the only thing people appropriately value, maximizing pleasure will not maximize well-being.

Second, within this constraint, trying to produce more of some values makes very good sense. Some values (pleasure and physical health, for example) are such that we should aim to increase them. Moreover, some values are so central to our systems of values that we ought to make producing more of them a priority. Physical health and leisure time seem to be two values like this. A lack of health

---

[16] Life satisfaction and positive affect do sometimes come apart. We can see this in studies of the relationship between income and well-being. See (Kahneman and Deaton 2010). For an example in which positive affect and rewardingness pull in different directions see White and Dolan (2009) on the relationship between having children and well-being.

[17] For more, see Tiberius (unpublished manuscript), and Raibley (2010).

makes the achievement of other values difficult: it's not easy to be a good parent, spouse, or friend when you have no energy; it's not easy to accomplish personal or career goals when you are coping with illness. Sufficient leisure time is similarly necessary for the pursuit of all sorts of values: a person who works 16 h a day does not have time to develop meaningful relationships or to pursue satisfying personal projects. The Value Fulfillment Theory would have us pay attention to what values are most important to people, to ask what are the necessary conditions for the achievement of these values, and to secure those conditions as much as we can. Thus, VFT favors a moderate and constrained form of promotion of the good that is sensitive to what people value.

Third, VFT points us to some things it makes sense to promote that we do not see by looking at other theories. In particular, agential capacities—the capacities to figure out what matters and to make choices about how to prioritize the various commitments one has—are crucially important to living a good human life. These capacities may not be appropriate targets of maximization, but it would make sense to promote the conditions under which they can be developed, which would include (among other things) effective education, political liberties, material resources and leisure.

5.5 Conclusion

I have argued that the only thing it makes sense to maximize is the all-encompassing good, and that, further, the only way such a thing could be maximized is if it can be meaningfully measured and quantified. I have also suggested that the human good—understood as well-being—is not like this. The human good includes many ingredients, some of which (such as friendship) are not appropriate targets for maximization. This way of understanding human well-being does also temper the goal of maximizing material wealth, though not by replacing "material wealth" with another good to be maximized. Instead, understanding human well-being in terms of the achievement of what we value leads us to think about what real good we can do by producing more wealth, or by maximizing material resources. Having more money is a good thing if it lets us achieve more of what we value. We should not ignore the fact that having more money does help people pursue their values, but money is not sufficient for the purpose, nor is it the end goal.

References

Annas J (1995) The morality of happiness. Oxford University Press, New York
Bentham J, Burns JH, Hart HLA (1996) An introduction to the principles of morals and legislation. Oxford University Press, Oxford
Brandt RB (1979) A theory of the good and the right. Oxford University Press, Oxford

Diener E, Scollon CN, Lucas RE (2003) The evolving concept of subjective well-being: the multifaceted nature of happiness. Adv Cell Aging Gerontol 15:187–219
Eid M, Larsen RJ (2008) The science of subjective well-being. Guilford Press, New York
Griffin J (1986) Well-being: its meaning, measurement, and moral importance. Clarendon Press, Oxford
Haybron DM (2008) The pursuit of unhappiness: the elusive psychology of well-being. Oxford University Press, Oxford
Heathwood C (2006) Desire, satisfactionism and hedonism. Philos Stud 128(3):539–563
Jaworska A (2007) Caring and internality. Res 74(3):529–568
Kahneman D, Deaton A (2010) High income improves evaluation of life but not emotional well-being. Proc Natl Acad Sci 107(38):16489–16493
Kahneman D, Krueger AB, Schkade DA, Schwarz N, Stone AA (2004) A survey method for characterizing daily life experience: the day reconstruction method. Science 306(5702):1776–1780
Keller S (2004) Welfare and the achievement of goals. Philos Stud 121(1):27–41
Kesebir P, Diener E (2008) In pursuit of happiness: empirical answers to philosophical questions. Perspect Psychol Sci 3(2):117–125
Mill JS (1979) Utilitarianism. Hackett Publishing Company (First published), Indianapolis, p 1863
Nozick R (1974) Anarchy, state, and utopia. Basic Books, New York
Raibley J (2010) Well-being and the priority of values. Soc Theor Pract 36(4):593–620
Railton P (1986) Moral realism. Philosophical Rev 95(2):163–207
Richardson HS (1997) Practical reasoning about final ends. Cambridge University Press, Cambridge
Rosati CS (1995a) Persons, perspectives, and full information accounts of the good. Ethics 105(2):296–325
Rosati CS (1995b) Naturalism, normativity, and the open question argument. Nous 29(1):46–70
Ryff CD, Singer B (1998) The contours of positive human health. Psychol Inquiry 9(1):1–28
Schmuck PE, Sheldon KM (2001) Life goals and well-being: towards a positive psychology of human striving. Hogrefe and Huber Publishers, Seattle
Seligman MEP (2002) Authentic happiness: using the new positive psychology to realize your potential for lasting fulfilment. Free Press, New York
Sumner L (1996) Welfare, happiness, and ethic. Clarendon Press, Oxford
Tiberius V (2008) The reflective life: living wisely with our limits. Oxford University Press, Oxford
Tiberius V, Plakias A (2010) Well-being. In: Doris J (ed) The moral psychology handbook. Oxford University Press, Oxford
Tiberius V "Well-Being for the Uninformed: Prudential Reasons and the Value Fulfillment Theory". unpublished
Velleman JD (1988) Brandt's definition of 'good'. Philos Rev 97(3):353–371
White MP, Dolan P (2009) Accounting for the richness of daily activities. Psychol Sci 20(8):1000–1008

Chapter 6
How Wise is Mother Nature? Maximization, Optimization and Short-Sighted Resource Use in Biological Evolution

Hanna Kokko

6.1 Introduction

The philosopher Daniel Dennett (1995) once stated that evolution is the best single idea that anyone has ever had. Of course, this statement does not refer to evolution as an invention, but as a discovery. Animals, plants, fungi and bacteria; the emergence of all of us is governed by the rules of natural selection, and evolutionary theory establishes the rules for understanding why the diversity of creatures look and behave the way they do.

When Charles Darwin published his work *On the Origin of Species* in 1859, its power to change the world was not the explicit statement that organisms evolve over time to become different. Educated people had already held such a view for a while. What Darwin's brilliant idea was really about was providing an explanation for adaptation that did not require any supernatural force. Organisms change over time, but this, according to Darwin, does not happen in any random direction, nor in a way that follows a predestined path; but all the same evolution creates traits that look deceptively as if someone had designed them to suit a particular purpose. The essential ingredient of Darwin's great idea is natural selection. It has the power to distinguish between useful and useless changes when random (and heritable) variations arise spontaneously in organisms. Therefore, organisms not only evolve, they also become better adapted over time.

Against this background, it may come as quite a surprise that biological evolution also churns out organisms that do not appear to maximize fitness, that fail to solve the problems set by the environment, or that act in ways that are detrimental to the performance of their populations. All these characteristics are, in fact, predicted by evolutionary theory, and they highlight a very central characteristic of 'Mother Nature'. Natural selection can definitely produce traits that appear to give

H. Kokko (✉)
Research School of Biology, ANU College of Medicine, Biology and Environment, Australian National University, Acton, Australia
e-mail: hanna.kokko@anu.edu.au

H. Brockmann and J. Delhey (eds.), *Human Happiness and the Pursuit of Maximization*, Happiness Studies Book Series, DOI: 10.1007/978-94-007-6609-9_6,

organisms 'foresight'—e.g., young migratory birds leaving their natal territories to fly south long before the first snow falls—but there are also mechanisms that guarantee that populations cannot be expected to always find solutions that appear 'wise' from the outside.

What does this have to do with happiness and maximization? Humans, as products of evolution, have aspirations that are, to a great degree, shaped by natural selection. It is therefore useful to know how, and in what sense, natural selection shapes organisms as 'maximizers'. Without necessarily knowing anything about the relevant psychological mechanisms (which, in sentient beings at least, are the driving forces that make an organism function in a particular way), it is of interest to know whether organisms indeed are designed, by natural selection, to maximize anything in particular. What will follow, in the examples below, is support for the idea that organisms might be built to strive towards something that might feel like happiness (of course, only often measurable through proxies, such as success in acquiring resources) while simultaneously these built-in mechanisms do not make it easy for evolved organisms to reach a level of resources where they are fully content with what they have. However, as we will see, the degree to which this is possible depends on the situation, most importantly whether or not one is dealing with a zero-sum game.

6.2 Why Not Reproduce as Fast as You Can?

At first glance one would assume that natural selection leads to every organism being maximally 'greedy'. A lot of success in a world dominated by natural selection, from the microbial realm to complex societies of insects or vertebrates, is determined by competitive interactions. All else being equal, faster reproduction wins. If a microbe processes organic matter faster and divides more often than its neighbors its genotype will dominate numerically later on. If a plant outgrows its neighbors, it will gather more sunlight and, all else being equal, produce more seeds. If a bird works harder to bring food back to the nest, it will probably be able to fledge more young than a neighbor that takes life a little more easily, maybe by not laying as many eggs in the first place.

Against such a background expectation, it can suddenly strike one as odd that any creature is ever found resting. Does life really favor the most frantic activities geared towards maximal reproductive output? And if it does not, what is wrong with our argument?

This was a question that fascinated David Lack, an ornithologist whose first claim to fame was in a study of Darwin's finches in the Galapagos archipelago, and who then in the 1940s and 50s turned his attention to the study of avian clutch size. Why do some bird species routinely lay three eggs per clutch, others four, or ten? Why does natural selection not always create birds that are efficient layers? Lack proposed that the egg number is optimized, not maximized (up to some physiological upper limit), and that such a view is fully compatible with natural selection (Kingsland 1995).

How come? Lack's answer was simple (Lack 1968): It does not pay to lay more eggs than a parent is likely to be able to feed, such that the offspring will fledge from the nest. If food availability ultimately limits the rate of food delivery to the nest, then parents who have laid more eggs than they can raise do not benefit from having attempted to maximize egg number. By 'not benefitting' an evolutionary biologist means that their genes are not disproportionately represented in future generations. Instead, clutch size is adjusted by natural selection to match the maximum number of nestlings the parents can feed. The number of surviving young is predicted to have a hump-shaped relationship to the number of eggs laid. It is therefore not surprising that parents can 'take it easy' during the egg laying stage. They will only work really hard for their young after these have hatched, which is a small fraction of their annual routine.

Lack made several predictions based on his idea. In high-latitude areas, food availability varies seasonally so much that large clutches can be profitably raised during the height of summer, while in the tropics year-round competition for food greatly limits its availability. In the tropics there is no point in time with super-abundant resources for feeding young, thus clutch sizes will be smaller.

While this much is true—clutch sizes indeed are larger in temperate areas than in the tropics—correlational data tend not to be enough to convince evolutionary biologists that a particular evolutionary explanation behind a pattern is responsible for creating it. It is often possible to ask nature much more direct questions by performing experiments. When Lack's optimal clutch size predictions were tested experimentally, it quickly emerged that not all outcomes fit the idea as neatly as expected.

Let's Test the Idea

A central prediction in Lack's theory was that offspring production should peak at the clutch size that is the norm in nature. Smaller as well as larger clutches should yield fewer offspring, though for different reasons: small clutches obviously cannot hatch many young, while young from large clutches face an increased risk of starvation. This prediction is quite easy to test experimentally, because birds often cannot determine the difference between eggs they have laid and any extra eggs that may have appeared there. This tendency to not question too much what one is incubating is what allows parasitic cuckoos to proliferate, but it also creates opportunities for curious researchers who can add or remove eggs laid by the appropriate species and see how many offspring the parents can raise to fledging.

The result? Perhaps surprisingly, many bird studies showed that parents actually can raise enlarged clutches to fledging. All they have to do is to work harder when there are more begging mouths in a nest, and indeed they seem to react this way. This means that Lack's idea cannot explain all there is to clutch size evolution. If a larger clutch was, after all, favored by natural selection, why did birds not become more frenetic layers over time?

More Refined Tests: Stable Isotopes Measure How Hard it is to be a Parent

A clue can be found in further experiments, exemplified by a study conducted in the mid-1990s by Serge Daan, Charlotte Deerenberg and Cor Dijkstra (Daan et al. 1996). Over a period of 6 years they located a total of 400 kestrel nests (a type of falcon), which they were able to pair in such a way that each pair of nests had similar laying dates and clutch size. Two young nestlings were moved from one nest to the other, which created an asymmetry between the nests: one set of parents was now able to work less hard, whilst the other had to hunt for mice and voles far more vigorously than before.

What did the researchers find now? The nests that had nestlings taken from them fledged, on average, around three young; the enlarged nests fledged more than six. Thus the extra young did not simply starve. The study instead repeated a finding already known for several other bird species: parents are able to feed more mouths adequately. In a new twist to the story, Daan et al. (1996) also employed a clever technique for estimating the daily energy expenditure of the parents: they analyzed the differential in clearance rates of stable isotopes of oxygen and hydrogen (the 'doubly labeled water method', Deerenberg et al. 1995). This provided a quantification for the conjecture that parents of enlarged broods must have been working harder: they did, by 22 %.

Would You Sacrifice Your Life for That of Your Child? A Kestrel-Based Calculation

Economically speaking, doubling the number of young raised by spending 22 % more energy might sound like a good deal. What is important, however, is whether and how the cost of energy is realized in the parent kestrel's body. Evolutionarily speaking, young newly fledged kestrels are a risky investment compared with the self-maintenance of a parent's body. A young bird of prey has a lot to learn before it is an efficient hunter, and many will die before reaching adulthood. An adult kestrel has already learned the tricks of the trade, and its reproductive value (formally quantified as its expected contribution to the future gene pool) is higher.

It is therefore important for an adult kestrel to keep its most valuable asset—its own body—in good enough shape. It is probably suboptimal to destroy all this capital in one go. Daan et al. (1996) tested this prediction from life history theory by showing that the 22 % increase in energy expenditure during brood raising had a near catastrophic effect on the probability that the parents survived the next winter. 60 % of birds who had been forced to raise two extra nestlings were reported dead before the end of the subsequent winter, while only 29 % of those raising control or reduced broods died (here control broods corresponded to natural brood sizes).

A back-of-the-envelope calculation shows what this means for kestrels. If we assume all mortality occurs during the winter, then 29 % annual mortality predicts an individual to be able to breed, on average, approximately 3.5 times. The higher mortality of 60 % predicts that birds, on average, manage to breed less than twice in their entire lifetime. Thus, would it make sense for kestrels to begin laying two extra eggs and work 22 % harder to get them fledge? Hardly, if this snaps off almost two entire breeding events from the parent's expected lifespan. If we additionally take into account that creating extra nestlings also requires laying the protein-rich eggs (rather than acquiring them for free, as in the experiment), the energy requirements of creating and dealing with extra nestlings would, obviously, grow larger still.

6.3 Kick Back and Relax? No, if It's a Zero-Sum Game

From the above, it might sound like evolution should equip us, like all animals, with quite a bit of good sense: overachievers are penalized and weeded out of the system. Unfortunately, it might be too rosy a view of life to expect that Mother Nature benevolently teaches individuals to relax in a way that prevents too frantic a life.

While the kestrel story is true, on its own it paints a biased picture of the rules of life. It is quite a typical example of *life-history theory* which predicts the scheduling of reproductive effort over an individual's lifetime (Stearns 1992). Within a life history context, optimization, not maximization, of effort often leads to maximal success. Bringing more offspring into the world is a task that the two parents attending a kestrel nest manage quite well, with much evidence of cooperation: while one parent hunts, the other keeps the chicks warm.

However, there are also plenty of cases where the interests of two individuals do not coincide, even if they are conspecifics. Parental effort is often fraught with conflict: each parent would benefit if the other worked a bit harder, unless there is lifelong monogamy. Perhaps surprisingly, this principle extends to the very definition of males and females: the sperm-producing individual can be seen to exploit the egg-producer's efforts (Lehtonen et al. 2012). Many aspects of reproduction are not captured by an optimization task, but rather by more game-theoretic notions of winning. Especially when we take into account that reproduction often involves both males and females, we find that surprisingly many situations in nature resemble a *zero-sum game* where one individual's gain is another individual's loss. When two male deer fight and the winner takes over the ownership of a harem of females, the situation is zero-sum. What does selection favor now?

Organisms are now, loosely speaking, not expected to follow any maxim such as 'our current level of investment is sufficient; it is silly to invest excessively in outcompeting others, let's all share resources more peacefully'.

6.4 When Does Evolution Optimize, When Does it Maximize?

There are, actually, at least three differences between zero-sum games and the situation that breeding kestrels find themselves in. The first is that when two or more individuals play games over finite resources, and the winner takes the lion's share of the resource (e.g., the unfertilized ripe eggs waiting inside a female deer), the best level of investment will now depend on one's rival's investment, featuring positive feedback. In other words, the larger the antlers and the more powerful the muscles of the opponent, the more energy you have to devote to developing yours to have any hope of gaining paternity. This situation is an *evolutionary arms race*, which—in simplistic models at least—often remains open-ended: it is not easy to see what would make the male half of the population stop developing ever more impressive weapons (Parker 2006).

Of course, in reality, the male reproductive effort too must at some point become optimized, rather than maximized: a deer that directs 100 % of its resources to antler growth will not survive, and a peacock that attempted to grow a kilometer-long tail would likewise be immobilized and quickly succumb to predation. Models that take the trade-off between survival and reproductive success into account often predict that males remain more modest in the development of their sexual extravaganza while they are still young (Kokko 1997). Still, the investment can as a whole evolve to be excessive because its evolution is governed by feedback from population-level investment, which has the effect of shifting the goalposts. The body size and level of armament required to outcompete rivals during some ancestral point in a species' history may be woefully inadequate in the present day.

When feeding kestrel chicks, the goalposts are not shifting to the same extent. The idea of shifting goalposts refers to the fact that what is good enough in an ancestral population will not be adequate once competitors have evolved new tricks (e.g., larger antlers). For the same idea to apply when parents feed their young, the same ability to capture and carry prey to the nest should be sufficient to raise chicks early in a species' history, but inadequate later. It is not obvious why this should be the case: the chicks themselves are selected to be able make good use of the prey delivered to the nest, they are thus not evolving to become ever more demanding.

Although there are also arms races between species, e.g., the chemical warfare between parasites and hosts, it is an intriguing thought that arms races are particularly common when conspecific individuals are trying to outcompete each other in reproductive success. The problem is disproportionately often a male one. Males of many species are larger as adults than the females they mate with, and they pay dearly for their faster growth as juveniles: they suffer higher mortality, especially in times of food shortage, and often have compromised immune systems compared with females who, loosely speaking, often appear to follow a more 'sensible' pattern of investment.

6.5 Group Optimality is Not Guaranteed

The other crucial difference in life-history optimization is a general feature of game theory: the behavior of the group is typically not optimized when the individuals maximize their own payoffs in response to others' actions. Again using deer as an example, there is no real sense in which the population benefits from half of its members devoting all available effort in attempts to outcompete each other. In some cases, like the northern elephant seal, males can become grotesquely large, weighing up to 2300 kg—more than three times as much as a female. One begins to wonder how much better the population would grow if males did not require so much food to grow to their large size, and if they did not harass females to the extent that pups can become crushed to death in the process.

Of course, we do not expect male deer, or male elephant seals, to manage to agree to share paternity with the available females in a peaceful manner. If there ever was a mutation that made an individual male behave as if its thinking obeyed the rule 'I've already mated with 3 different females in my lifetime; surely this is an achievement I can be fully content with—I can retire now', it is not hard to see that this male's genes will not be passed on to the next generation as often as those of his more fervently competitive rivals. A mentality where nothing is ever enough will prevail.

6.6 Sexual Conflict: Males Can Do a Lot of Damage

The third intriguing feature, which is specific to sexually reproducing species with separate males and females, is that the female part of the population, i.e. that actually responsible for physically producing the new generation, can suffer detrimental consequences when selection acts on males (Connallon et al. 2010). In cases where populations are managed for economic or recreational use, e.g., hunting, it is often customary to remove a disproportionate number of males. When there are fewer males eating away at the grass (deer) or fish (seals), there will logically be more food available for the females, who can then produce the next generation more efficiently. In experiments, it is possible to create the opposite scenario with male-biased enclosed populations, and in some species such experiments reveal terrible consequences: when female common lizards *Lacerta vivipara* are in the minority of a local population they can become sexually harassed to such a degree that they suffer bites and reduced time left for foraging, which then manifests itself as higher mortality and lower fecundity (Le Galliard et al. 2005). When females become more short-lived, male attention per living female automatically increases (as there are now more males per female), and the vicious circle can end in the extinction of the entire local population (Le Galliard et al. 2005; Rankin et al. 2011). Milder forms of the same conflict can be found e.g., in guppies, where females accept riskier, predator-infested feeding areas to escape unwarranted male attention (Darden and Croft 2008).

Sexual conflict can have harmful effects even when it is not expressed in any dramatic behaviors at a behavioral level. In bank voles, males with high testosterone (T) have higher mating success, and they pass on these high-T genes to their offspring. While this elevates the mating success of their sons, these same genes, when inherited by the daughters, lower the fecundity of these females (Mokkonen et al. 2011). This is an example of *intralocus* sexual conflict, which arises due to the fact that males and females share much of their genomes, yet face different selection pressures. When one sex—typically males—is selected to favor fast success at the expense of others, similar traits can arise in the sex that does not benefit from such behavior. The situation depicts a confused Mother Nature, facing a tug-of-war between effects that are beneficial to males and detrimental to females; undecided she goes for an intermediate solution that is not ideal for either sex.

It's Not Just About Sex

Selection that differs between the sexes is not the only example where population performance suffers because individuals shift some of their investment 'portfolio' from traits that are good for individuals and populations alike (e.g., the production of eggs, care of young), towards traits that simply function to outcompete conspecifics. Zero-sum games can arise in many contexts where a set of individuals is competing for a finite resource, and the first one to secure a quick win has an improved payoff while others suffer. It is not too far-fetched to draw attention to similarities between a process in which males drive females to ever lower numbers—by biting, by forcing them to feed in poorer areas, or by virtue of growing so large that there is little left for females—and the greed with which humans overexploit the oceans for food.

A female's eggs that are fertilized by male A are unavailable to male B, just like a tuna that is caught by a Spanish vessel is unavailable to be caught by a Scottish one. This selects behaviors that benefit individuals in the short term (get the prize ahead of others), without much regard for the long-term benefits (i.e. ensuring prizes are maintained such that future generations can enjoy them). As soon as the technology existed to seriously overexploit the planet's fish stocks, the decline in this globally rich source of protein began in earnest (Clover 2006;

Ludwig et al. 1993). Negotiations to limit the damage to fish populations are, unsurprisingly, always fraught with difficulty.

And Even if Sex Causes Trouble, You Don't Have to Be a Male to Be Shortsighted

The shortsightedness of Mother Nature is, ultimately, an expected outcome of evolutionary processes because evolution never plans ahead. It rewards strategies that have worked well in the past, and because each individual is genetically rewarded if it gets 'more than the average' (as opposed to the 'maximum possible' which might require skillful cooperation with conspecifics), it does not easily find solutions that are good for the species in the long term. Above, I have blamed males for much of the trouble they cause, and there is a good scientific reason to do so: the very definition of a male is related to the difference in gamete size, with males producing small sperm and females large eggs. In species where the nutrition contained in gametes is the only form of parental investment—and this is the majority of all species—the very definition of a male implies that this sex evolves to parasitize female reproductive effort in a way that diminishes population growth (Lehtonen et al. 2012).

However, this does not mean that the other sex is immune to the detrimental effects of evolution's shortsightedness. A perhaps extreme case is provided by Amazon mollies *Poecilia formosa*, which is a small fish that turns out to never produce any males at all (Schlupp 2009). This all-female species arose as a hybrid between two other *Poecilia* molly species, and its weirdest characteristic is that its eggs still need contact with sperm to begin their development. Genes from males are not incorporated in the genome, but the rule, established during evolutionary history that tells eggs to wait until sperm touches them is still obeyed.

The asexual Amazon molly can therefore only reproduce when males of another *Poecilia* species are present. Genetic incompatibilities are not a problem here, because the offspring will be a clone of its mother. But this weird reproductive system, called *gynogenesis* or *sperm-dependent parthenogenesis* (Beukeboom and Vrijenhoek 1998), brings about plenty of other problems. Because Amazon females do not waste any effort producing males, an Amazon female can afford twice the number of daughters compared with a female of the other *Poecilia* species. Each of these females can again produce twice the number of daughters compared with other species, and by now the prediction is clear: the species that still produce males starts to be in short supply. Just like female lizards formed a resource that was eventually in short supply in the examples above, now male Poecilia suffer the same fate. While they (unlike the female lizards) are not physically harmed during mating, they are simply produced in dwindling numbers because Amazon females outcompete the females of the sexual *Poecilia* species (Kokko and Heubel 2011).

If Amazon females had foresight, they would reduce their fecundity below that of the sexual *Poecilia* species. Why? Their 'unfair' advantage of producing female offspring only, while 'borrowing' the males of another species for their sperm, allows them to outcompete other *Poecilia* species that they ultimately depend on for a sperm supply. Once the other species have been driven to extinction, there is suddenly no sperm for Amazons either, and they also become extinct. The molly system can keep on existing, however, because by the time a local extinction happens some mollies may have moved to other parts of a larger water system (Kokko et al. 2008).

6.7 What Do Studies on Mollies and Lizards Really Tell Us?

The lesson? Female mollies are no better than male lizards when it comes to keeping a resource intact for future generations. In both cases, the evolution of self-restraint is unlikely because the problem (lack of males, or lack of females, respectively) does not arise immediately but is felt more strongly by future generations. Simultaneously, there is the problem that one individual cannot, by changing its own behavior, change much of the behavior of the entire population. If it showed restraint in resource use, others would carry on depleting this.

In the case of the lizard, we could imagine a mutation in a male that makes it refrain from the most aggressive ways to grab females. Female lizards would still have a hard time with the other males, and the 'kind' male would simply lose out in the number of copulations—and his genes will not be passed on. Similarly, Amazon mollies showing self-restraint would give birth to fewer offspring (here it is good to remind ourselves that this language does not imply conscious decision-making; we are dealing with a hypothetical mutation that lowers the fecundity of a female, perhaps by changing the number of follicles developing in the ovaries). Lowering the fecundity would make the genes of this mother less prevalent in future generations, while others would still march ahead and reproduce rapaciously.

This is why group-beneficial traits that involve restraint have a difficult time evolving: the individual showing restraint is often the one who is penalized in terms of the number of gene copies it transmits to future generations. The view that evolution always favors traits that enable a species to persist is therefore rather naïve. Instead, natural selection in the short term can make individuals maximize mating success or fecundity, in a way that short-changes future generations.

6.8 Conclusion

It takes great optimism to draw an analogy between natural selection and the idea of an 'invisible hand' that somehow makes sure that individual ambition to reach one's own goals benefits society in general. Unfortunately, current biological knowledge shows that individual optimization is in no way guaranteed to lead to group-optimal outcomes.

All this, of course, forms a sobering message to humans too. There are many ways in which humans are unique, and when dealing with resources we all depend on, one crucial difference to the non-human world is our ability to agree on social norms that create better solutions to problems than would be achievable through letting everyone behave as they please. Simultaneously, however, the discussions aimed at creating norms show signs of an underlying selfishness and short-sightedness. Otherwise it would not be so difficult to agree on cutting quotas when fish stocks show signs of becoming unsustainably depleted, or cutting CO_2 emissions when climate change threatens the wellbeing of entire generations in the future.

Interestingly, however, it appears that humans in the developed world have shifted from maximizing to optimizing offspring number, which in itself is an interesting experiment in life history evolution (Lawson and Mace 2011). The hunger for more wealth, however, appears surprisingly insatiable, to the great detriment of the environment. Bill Hamilton's (1971) statement that the animal part of our nature is greatly concerned with getting 'more than the average', of course predicts exactly that; we are not that much better than male lizards when it comes to dealing with zero-sum games.

References

Beukeboom LW, Vrijenhoek RC (1998) Evolutionary genetics and ecology of sperm-dependent parthenogenesis. J Evol Biol 11:755–782

Clover C (2006) The end of the line: how over-fishing is changing the world and what we eat. The New Press, New York

Connallon T, Cox RM, Calsbeek R (2010) Fitness consequences of sex-specific selection. Evolution 64:1671–1682

Daan S, Deerenberg C, Dijkstra C (1996) Increased daily work precipitates natural death in the kestrel. J Anim Ecol 65:539–544

Darden SK, Croft DP (2008) Male harassment drives females to alter habitat use and leads to segregation of the sexes. Biol Lett 4:449–451

Darwin C (1859) On the origin of species by means of natural selection. Murray, London

Deerenberg C, Pen I, Dijkstra C, Arkies BJ, Visser GH, Daan S (1995) Parental energy expenditure in relation to manipulated brood size in the European kestrel Falco tinnunculus. Zool Anal Complex Syst 99:38–47

Dennett DC (1995) Darwin's dangerous idea. Touchstone Books, New York

Hamilton WD (1971) Selection of selfish and altruistic behaviour in some extreme models. In: Eisenberg JF, Dillon WS (eds) Man and beast: comparative social behaviour. Smithsonian Press, Washington, DC, pp 57–91

Kingsland SE (1995) Modeling nature. University of Chicago Press, Chicago
Kokko H (1997) Evolutionarily stable strategies of age-dependent sexual advertisement. Behav Ecol Sociobiol 41:99–107
Kokko H, Heubel KU (2011) Prudent males, group adaptation, and the tragedy of the commons. Oikos 120:641–656
Kokko H, Heubel KU, Rankin DJ (2008) How populations persist when asexuality requires sex: the spatial dynamics of coping with sperm parasites. Proc R Soc Lond B 275:817–825
Lack D (1968) Ecological adaptations for breeding in birds. Chapman & Hall, London
Lawson DW, Mace R (2011) Parental investment and the optimization of human family size. Philos Trans R Soc Lond B 366:333–343
Le Galliard J-F, Fitze PS, Ferrière R, Clobert J (2005) Sex ratio bias, male aggression, and population collapse in lizards. Proc Nat Acad Sci USA 102:18231–18236
Lehtonen J, Jennions MD, Kokko H (2012) The many costs of sex. Trends Ecol Evol 27:172–178
Ludwig D, Hilborn R, Walters CJ (1993) Uncertainty, resource exploitation and conservation: lessons from history. Science 260:17–36
Mokkonen M, Kokko H, Koskela E, Lehtonen J, Mappes T, Martiskainen H, Mills SC (2011) Negative frequency dependent selection maintains sexually antagonistic alleles in "Myodes glareolus". Science 334:972–974
Parker G (2006) Sexual conflict over mating and fertilization: an overview. Philos Trans R Soc B 361:235–259
Rankin DJ, Dieckmann U, Kokko H (2011) Sexual conflict and the tragedy of the commons. Am Nat 177:780–791
Schlupp I (2009) Behaviour of fishes in the sexual/unisexual mating system of the Amazon molly (Poecilia formosa). Adv Study Behav 39:153–183
Stearns SC (1992) The evolution of life histories. Oxford University Press, Oxford

Chapter 7
Towards a Neuroscience of Well-Being: Implications of Insights from Pleasure Research

Kent C. Berridge and Morten L. Kringelbach

7.1 Introduction

The study of well-being or positive psychology is part of a long tradition reaching back to Aristotle, where well-being or happiness has been usefully proposed to consist of at least two ingredients: *hedonia* and *eudaimonia* (Aristotle 350 B. C. 2009; Seligman et al. 2005). Definitions by philosophers and psychologists have varied, but most generally agree that hedonia corresponds psychologically to pleasure. By comparison, eudaimonia has been less easy to define, but for most it corresponds to some aspect of a life well lived and not to any particular emotional state. In this review, we take eudaimonia to mean essentially a life experienced as valuably meaningful and as engaging.

Hedonic processing and eudaimonic meaningfulness may thus appear very different in terms of definition and conceptualization. At the same time, empirical findings have been found well-being to involve both together. Questionnaire scores for hedonia and eudaimonia typically converge in the same individuals (Diener et al. 2008; Kuppens et al. 2008). Thus, if a person self-reports to be hedonically happy, then that same person is also likely to report a high sense of positive meaningfulness in life.

The tendency for coherence between ratings of pleasure and meaningfulness opens a potential window of opportunity for the neuroscientific study of both

The authors contributed equally to this work.

K. C. Berridge (✉)
Affective Neuroscience and Biopsychology Lab, Department of Psychology,
University of Michigan, Ann Arbor, MI, USA
e-mail: berridge@umich.edu

M. L. Kringelbach
Department of Psychiatry, University of Oxford, Oxford, UK
e-mail: morten.kringelbach@psych.ox.ac.uk

H. Brockmann and J. Delhey (eds.), *Human Happiness and the Pursuit of Maximization*, Happiness Studies Book Series, DOI: 10.1007/978-94-007-6609-9_7,

aspects of well-being (Kringelbach and Berridge 2009; Urry et al. 2004). If both ingredients occur in the same people, then the neurobiological bases for both coexist in the same brains. If both cohere, then identifying neural markers of one may give a toehold into identifying the other. Still, most would probably agree that eudaimonic happiness poses harder challenges to psychology and neuroscience. It is difficult even to define life meaningfulness in a way as to avoid dispute, let alone to tie a happy sense of meaningfulness to any specific brain patterns of activation. The difficulties of approaching eudaimonic meaning are not insurmountable in principle, but for the foreseeable short term seem likely to remain obstacles to affective neuroscience.

We have therefore chosen to focus mostly upon the hedonia or pleasure aspect of well-being. The pleasure aspect is far more tractable, and can be inspected against a growing background of understanding of the neural foundations for specific pleasures. Supporting a hedonic approach to happiness, happy people typically take more pleasure from life. Indeed it has been suggested that the best and simplest measure of well-being may be to merely ask people how they hedonically feel right now—again and again—so as to track their hedonic accumulation across daily life (Kahneman 1999a). Such repeated self-reports of hedonic *states* could also be used to identify more stable neurobiological hedonic brain *traits* that dispose particular individuals toward happiness.

Conversely, most will agree that the capacity for pleasure is essential to normal well-being. The pathological loss of pleasure, anhedonia, which is found in many affective disorders is devastating and precludes well-being. Our aim in this review is to highlight findings from recent research on brain mechanisms of pleasure and to ask how to higher states of hedonia might be generated to produce well-being, and conversely what might go wrong in affective disorders (Berridge and Kringelbach 2008; Kringelbach and Berridge 2010b; Leknes and Tracey 2010; Smith et al. 2010).

In passing, we note that our focus on the hedonia component of happiness should not be confused with hedonism, which is the pursuit of pleasure for pleasure's own sake, and more akin to the addiction features we describe below, which does not necessarily involve much actual pleasure. We also note that while our focus is mainly on mechanisms of stimulus-bound sensory pleasure, this reflects merely current experimental research, and the evidence appears to show that pleasure generators can be independent of sensory input as found, for example, in locked-in patients (Bruno et al. 2011). Further, to focus on hedonics does not deny that some ascetics may have found bliss through painful self-sacrifice, but simply reflects that positive hedonic tone is indispensable to most people seeking happiness (Diener et al. 2008; Gilbert 2006; Kahneman 1999a; Seligman et al. 2005).

7.2 A Science of Pleasure

Pleasure has been proposed to be evolution's boldest trick allowing species and organisms to ensure survival and procreation in both individuals and species (Kringelbach 2009). Substantial mechanisms for pleasure would be selected for and conserved only if they ultimately served a central role in fulfilling Darwinian imperatives of gene proliferation via improved survival and procreation, suggesting the capacity for pleasure must have been fundamentally important in evolutionary fitness (Cabanac 2010; Darwin 1872; Nesse 2002; Panksepp 1998).

Pleasure is never merely a sensation, even for sensory pleasures (Frijda 2010; Kringelbach 2010; Kringelbach and Berridge 2010b; Ryle 1954). Instead pleasure always requires the recruitment of specialized brain systems to actively paint an additional "hedonic gloss" onto a sensation. Active recruitment of brain pleasure-generating systems is what makes a pleasant experience 'liked' (Fig. 7.1).

The capacity of certain stimuli, such as a sweet taste or a loved one, to reliably elicit pleasure—to nearly always be painted with a hedonic gloss—reflects the privileged ability of such stimuli to engage these hedonic brain systems responsible for manufacturing and applying the gloss. Hedonic brain systems are well-developed in the brain, spanning subcortical and cortical levels, and are quite similar across humans and other animals.

Some might be surprised by high similarity across species, or by substantial subcortical contributions, at least if one thinks of pleasure as uniquely human and as emerging only at the top of the brain. The neural similarity indicates an early phylogenetic appearance of neural circuits for pleasure and a conservation of those circuits, including deep brain circuits, in the elaboration of later species, including humans.

The fundamental rewards afforded by biological evolution include food, sex and conspecifics. Food is one of the most universal routes to pleasure (Kringelbach 2004). Sex is another potent natural sensory pleasure which involves some of the same brain circuits (Georgiadis and Kringelbach 2012). Many other special classes of stimuli also appear tap into the same limbic circuits (Everitt et al. 2008; Kelley and Berridge 2002; Koob and Volkow 2010).

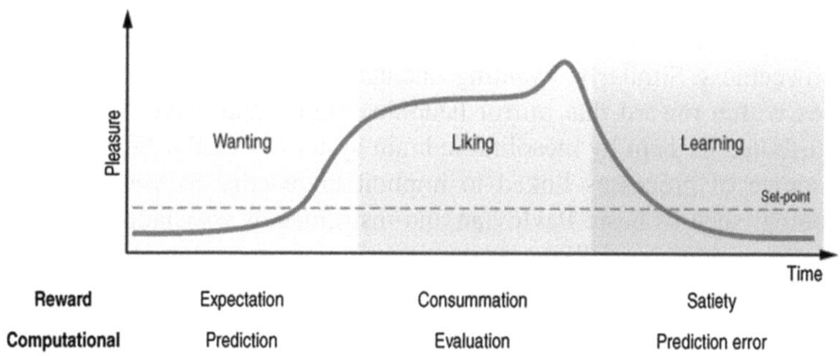

Fig. 7.1 Pleasure cycles

Also social interaction with conspecifics draws on overlapping neural systems (Frith and Frith 1999). In fact, it might well be even from an evolutionary perspective that in humans, at least, the social pleasures are often as pleasurable as the basic sensory pleasures.

Most uniquely, humans have many prominent higher order, abstract or cultural pleasures, including personal achievement as well as intellectual, artistic, musical, altruistic, and transcendent pleasures. While the neuroscience of higher pleasures is in relative infancy, even here there seems overlap in brain circuits with more basic hedonic pleasures (Frijda 2010; Harris et al. 2009; Leknes and Tracey 2010; Salimpoor et al. 2011; Skov 2010; Vuust and Kringelbach 2010). As such, brains may be viewed as having conserved and re-cycled some of the same neural mechanisms of hedonic generation for higher pleasures that originated early in evolution for simpler sensory pleasures.

7.3 The Neuroanatomy of Pleasure and Reward

Our subjective experience may suggest that a state of positive affect is a unitary process, but affective neuroscience analyses have indicated that even the simplest pleasant experience, such as a mere sensory reward, is actually a more complex set of cyclical processes containing several psychological components, each with distinguishable neurobiological mechanisms (Berridge et al. 2009; Kringelbach and Berridge 2009; Leknes and Tracey 2010). These include at least the three components of wanting, liking and learning. *Liking* is the actual pleasure component or hedonic impact of a reward, *wanting* is the motivation for reward and *learning* includes the associations, representations and predictions about future rewards based on past experiences (Fig. 7.2).

We distinguish between the conscious and non-conscious aspects of these subcomponents. Both exist in people (Winkielman et al. 2005), but the latter at least can also be studied in other animals in ways that help better reveal the underlying neural generating mechanisms. At the potentially non-conscious level, we use quotation marks to indicate that we are describing objective, behavioural or neural measures of these underlying brain processes. As such, 'liking' reactions result from activity in identifiable brain systems that paint hedonic value on a sensation such as sweetness. Similarly, 'wanting' includes incentive salience or motivational processes within reward that mirror hedonic 'liking' and make stimuli attractive when attributed to them by mesolimbic brain systems. Finally, 'learning' includes a wide range of processes linked to implicit knowledge as well as associative conditioning, such as basic Pavlovian and instrumental associations.

At the conscious level liking is the conscious experiences of pleasure, in the ordinary sense of the word, which may be elaborated out of core 'liking' reactions by cognitive brain mechanisms of awareness. Conscious wanting includes conscious desires for incentives or cognitive goals, while learning includes the updating of explicit and cognitive predictions (Friston and Kiebel 2009; Zhang et al. 2009).

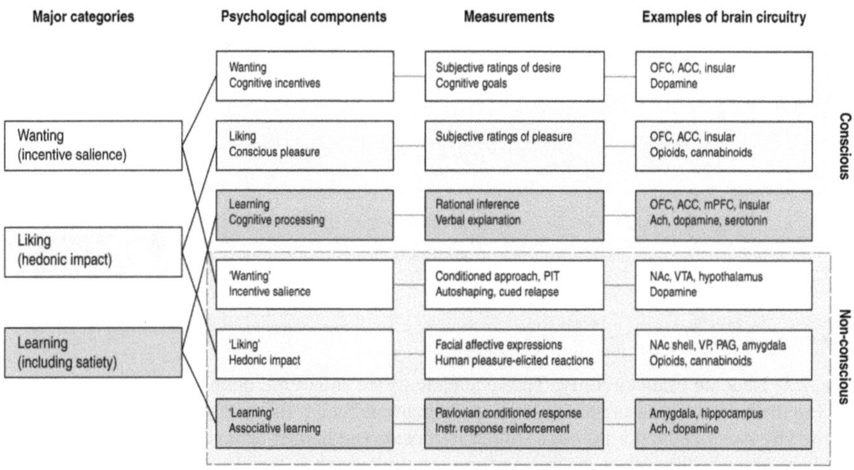

Fig. 7.2 Measuring reward and hedonia. Hedonic reward processes related to well-being are multifaceted psychological concepts that constantly interact and require careful scientific analysis to tease apart. Measurements or behavioral procedures that are especially sensitive markers of the each of the processes are listed (*third column*)

This universal experience of pleasure as a consciously felt feeling is perhaps the reason why pleasure has seemed purely subjective to many thinkers. But related to the notion that pleasure naturally evolved, we suggest that pleasure also has objective aspects that can be detected in brain and mind. Note again, however, the underlying similarities of brain mechanisms for generating sensory pleasures in the brains of most mammals, both humans and nonhumans alike (Fig. 7.3). It seems unlikely so much neural machinery would have been selected and conserved across species if it had no function. Basic pleasure reactions have always had objective consequences, and brain mechanisms for hedonic reactions have long been functionally useful—even before any additional mechanisms appeared that characterize any human-unique aspects of subjective feelings of pleasure. In a sense, we suggest hedonic reactions have been too important to survival for pleasure to be exclusively subjective.

Pleasure Generators: Hedonic Hotspots in the Brain

How does pleasure actually arise in a brain? The brain appears frugal in mechanisms that are causally sufficient to generate 'liking' or magnify pleasure to high levels. These few mechanisms are candidate brain wellsprings for hedonic happiness.

Compelling evidence for pleasure causation as increases in 'liking' reactions has so far been found for only a few brain substrates, or hedonic hotspots. Those hedonic hotspots mostly reside—surprisingly, if one thought pleasure to be

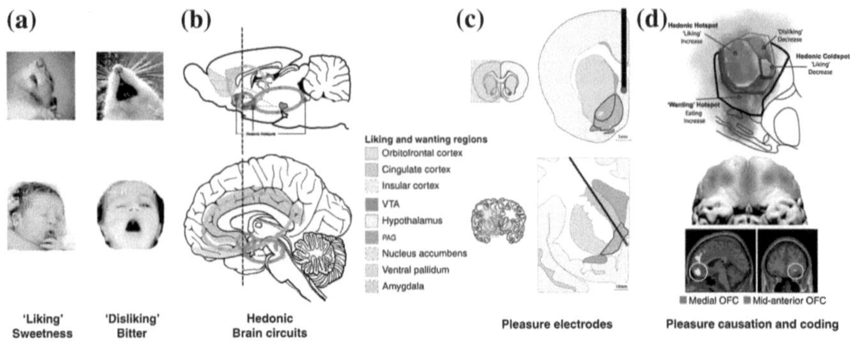

Fig. 7.3 Hedonic brain circuitry. The schematic figure shows the brain regions for causing and coding fundamental pleasures in rodents and humans. **a** Facial 'liking' and 'disliking' expressions elicited by sweet and bitter taste are similar in rodents and human infants. **b, d** Pleasure causation has been identified in rodents as arising from interlinked subcortical hedonic hotspots, such as in nucleus accumbens and ventral pallidum, where neural activation may increase 'liking' expressions to sweetness. Similar pleasure coding and incentive salience networks have also been identified in humans. **c** We believe the so-called 'pleasure' electrodes in rodents and humans were unlikely to have elicited much true pleasure but perhaps only incentive salience or 'wanting'. **d** The cortical localization of pleasure coding may reach an apex in various regions of the orbitofrontal cortex, which differentiate subjective pleasantness from valence processing of aspects the same stimulus, such as a pleasant food

cortical—deep below the neocortex in subcortical structures. Our strategy to find such neural generators of pleasure gloss has relied on activating neural mechanisms underlying natural 'liking' reactions to intensely pleasant sensations. An example of 'liking' is the positive affective facial expression elicited by the hedonic impact of sweet tastes in newborn human infants (Fig. 7.3a), such as tongue protrusions that can lick the lips. By contrast, nasty bitter tastes instead elicit facial 'disliking' expressions of disgust such as gapes, nose and brow wrinkling, and shaking of the head. Many of these affective expressions are similar and homologous (sharing features such as identical allometric timing laws) in humans, orangutans, chimpanzees, monkeys, and even rats and mice (Steiner et al. 2001). Homology in origin of 'liking' reactions implies that the underlying hedonic brain mechanisms are similar in humans and other animals, opening the way for an affective neuroscience of pleasure generators that bridges both.

Subcortical Hedonic Hotspots in Nucleus Accumbens, Ventral Pallidum and Brainstem

Some insight into pleasure-causing circuitry of human brains has been gained by affective neuroscience studies in rodents in which the hedonic hotspots are activated to magnify a sensory pleasure, and so reveal the location and neurotransmitter identity of the generating mechanism for intense 'liking'. A hedonic hotspot is capable of generating enhancements of 'liking' reactions to a sensory pleasure such as sweetness, when opioid, endocannabinoid or other hedonic neurochemical circuits within the hotspot are stimulated (Mahler et al. 2007; Peciña and Berridge 2005; Peciña et al. 2006; Smith and Berridge 2005). In rodent studies, the hotspots can be activated by painless microinjections of drug droplets that stimulate neurotransmitter receptors on neurons nearby. Within the hotspot, drug microinjections magnify the hedonic impact of a sweet pleasure, whereas outside the border of the hotspot the same microinjections fail to elevate 'liking'.

The results of such studies reveal a network of brain hedonic hotspots, distributed as a chain of 'liking'-enhancing islands of brain tissue across several deep structures of the brain. The network of hedonic hotspots acts together as a coordinated whole to amplify core pleasure reactions. Each brain hotspot may be merely a cubic-millimeter or so in volume in the rodent brain (and would be expected to be a cubic-centimeter or so in you, if proportional to the larger human volume of whole brain). The small size of each anatomical hotspot indicates a surprisingly localized concentration of sufficient-cause mechanisms for generating an intense pleasure in the brain. The network properties reveal a fragile substrate for pleasure enhancement that requires unanimity across the several parts in order to elevate hedonic 'liking' (Peciña 2008; Peciña and Smith 2010; Smith et al. 2010).

A major hotspot has been found in the nucleus accumbens, a brain structure at the bottom font of the brain, specifically in its medial shell region near the center of the structure. Other hotspots have been found further back in the brain. For example, a very important hedonic hotspot lies in the ventral pallidum, which is near the hypothalamus near the very bottom center of the forebrain and receives most outputs from the nucleus accumbens. Still other hotspots may be found in more distant parts of the rodent brain, possibly as far front in limbic regions of prefrontal cortex, and almost certainly as far back as deep brainstem regions including the parabrachial nucleus in the top of the pons.

Analogous to scattered islands that form a single archipelago, the network of distributed hedonic hotspots forms a functional integrated circuit, which obeys control rules that are largely hierarchical and organized into brain levels (Aldridge et al. 1993; Berridge and Fentress 1986; Grill and Norgren 1978; Peciña et al. 2006). At the highest levels, the hotspot network may function as a more democratic heterarchy, in which unanimity of positive votes across hotspots is required in order to generate a greater pleasure. For example, any successful enhancement that starts in one hotspot involves recruiting neuronal activation across other hotspots simultaneously, to create a network of several that all vote 'yes' together for more pleasure. Conversely, a pleasure enhancement initiated by opioid activation of one hotspot can be vetoed by an opposite vote of 'no' from another hotspot where opioid signals are suppressed. Such findings reveal the need for unanimity across hotspots in order for a greater pleasure to be produced, and the potential fragility of hedonic enhancement if any hotspot defects (Smith and Berridge 2007; Smith et al. 2010).

But all of these findings on brain pleasure generators are focused on making pleasures *nicer than usual*. Neurochemical activation of hedonic hotspots creates a brain wellspring for intense pleasure when candidate sensations are encountered, generating high hedonic peaks of sensory pleasure.

Yet well-being is a more continuous and quotidian state of *hedonic normalcy* in a slightly positive balance. What in the brain is required for creating the daily continual baseline level of a normal pleasure gloss? It turns out that only some of the hotspots that amplify pleasure are necessary for normal hedonic levels of 'liking' to pleasant sensations, and particularly the one in ventral pallidum.

In both the clinical literature and in our experiments, normal core 'liking' reactions to pleasure are relatively difficult to abolish absolutely by any single event, condition, brain lesion or drug (Bruno et al. 2011; Peciña 2008; Peciña and Smith 2010; Smith et al. 2010). Resilience of brain circuits for normal baseline pleasures may be very good in evolutionary terms.

Hedonic resilience may also be related to why many people can eventually regain a reasonably happy state even after catastrophic events (Diener et al. 2006; Gilbert 2006; Kahneman 1999b). Strikingly, hedonic balance may be retained even in the most extreme situations. One of the most extreme situations must surely be locked-in syndrome, a brain condition that leaves the person fully aware and cognitively intact but completely paralyzed to the extent of being able only to make slight movements of an eye or eyelid. Yet in the face of even this devastating degree of paralysis, locked-in patients may often still be happy. A recent study found that 72 % of locked-in respondents did report themselves to be moderately happy. The average response of this happy yet massively incapacitated group was +3 out of a hedonic scale from −5 to +5, where +3 corresponded to 'very well' (between +2 = 'well', and +4 = "almost as well at the best period in my life prior to having locked-in syndrome"). The remaining 28 % of locked-in respondents, who were much more likely to also be experiencing pain, reported themselves to be unhappy at −4, but even this corresponded only to "almost as bad as the worst period in my life before locked-in syndrome" (and not quite as bad as −5 = "as bad as the worst period in my life before"); only 7 % wished for euthanasia (Bruno et al. 2011). Hedonic resilience can apparently often persist with seemingly little to go on, still generated by hedonic circuits within the person.

Perhaps not surprisingly then, only one brain lesion has been found in our lab studies to destroy a normal sensory pleasure, and convert sweetness into a nasty experience: the ventral pallidum hotspot. This site is still preserved in locked-in patients, perhaps contributing to their remaining well-being. Damage to this unique brain site abolishes hedonic 'liking' reactions to sweetness and replaces instead with disgust or 'disliking' reactions (e.g., gapes) as though the sweet taste had turned bitter (Berridge et al. 2010; Cromwell and Berridge 1993; Smith et al. 2010). The ventral pallidum is the chief recipient of output from the nucleus accumbens and part of a corticolimbic circuit that extends from prefrontal cortex to nucleus accumbens to ventral pallidum, before looping up via thalamus to begin the circuit all over again in prefrontal cortex (Smith et al. 2010).

An important question is how similar or different the ventral pallidum role in pleasure might be in humans compared to in rodents. Currently we do not have much available data on the hedonic consequences of human hotspot loss, because a human stroke or tumor lesion rarely damages the ventral pallidum on both sides of the brain without also damaging hypothalamus and related structures in between, producing severe incapacitation that precludes much psychological assessment. Yet, in a rare human case report of a brain lesion that did rather selectively damage the ventral pallidal region on both sides without much else, positive affect and craving for rewards was reported to be much reduced. The patient's brain had incurred damage to ventral pallidum (and nearby medial globus pallidus) due to oxygen starvation when the patient stopped breathing during an enormous drug (Miller et al. 2006). Afterwards the pallidal-lesion patient reported that his feelings became dominated by depression, hopelessness, guilt, and anhedonia. Even formerly craved and hedonic sensations like drinking alcohol lost their feelings of pleasure for him, and he no longer craved the many drugs of abuse that he had previously avidly consumed. Even this lesion probably did not fully destroy his ventral pallidum, and perhaps this is why he was not as strongly seized by disgust as a rat would be if it had complete lesions of the ventral pallidum hotspot. Instead, the patient still continued to eat and drink normally after his lesion, and even gained weight. But his apparent dramatic decline in hedonic well-being suggests impairment in normal pleasure, and helps confirm a continuity between the ventral pallidum hotspot and human hedonia. We have also encountered anecdotal evidence that in some patients with pallidotomies (of nearby globus pallidus, just above and behind the human ventral pallidum) for Parkinson's patients, this led to severely flattened affect or anhedonia (Aziz, personal communication).

The striking restriction of brain substrates where damage converts 'liking' to 'disliking' seems a testimonial to the robustness of the brain's capacity for a basic pleasure reaction, and also perhaps an insight into what pathological mechanisms result in true anhedonia.

Additional Pleasure Codes in the Brain

The occurrence of pleasure is coded by neural activity in many additional forebrain sites beyond the hotspots mentioned above, including in amygdala and in the cortex: especially prefrontal cortical regions such as orbitofrontal cortex, anterior cingulate cortex, and insular cortex, (Grabenhorst and Rolls 2011; Kringelbach 2010; Salimpoor et al. 2011) (Fig. 7.4).

But not all brain structures that *code* for pleasure actually help to *cause* it. *Coding* of pleasure in the brain can reflect not only pleasure causation but also the neural consequences of pleasure: brain activity that results from pleasure enhancement but causes another function, such as cognition or learning. This implies that some brain activity may both cause and code pleasure reactions, whereas others do not cause pleasure but may code it while causing other

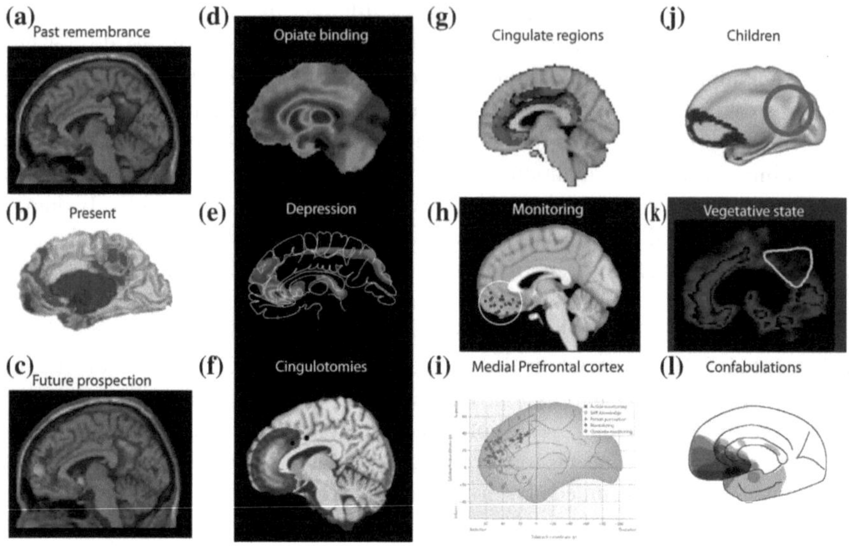

Fig. 7.4 The brain's default network and eudaimonic—hedonic interaction. **a–c** The brain's default network has been linked to self awareness, remembering the past and prospecting the future. Some components overlap with pleasure networks. We wonder whether happiness might include a role for the default network, or for related neural circuits that contribute to computing relations between self and others, in evaluating eudaimonic meaning and interacting with hedonic circuits of positive affect. Examples show key regions of the default network such as **d** the anterior cingulate and orbitofrontal cortices that have a high density of opiate receptors, **e** have been linked to depression, and **f** its surgical treatment **g** have been implicated by connectivity analyses, **h** are implicated in pleasure-related cognitive functions such as monitoring, learning and memory, **i** or in self-knowledge, person perception and other cognitive functions. **j** The default network may change over early life in infants and children, **k** in pathological states including depression and vegetative states, **l** and after cortical lesions that disrupt reality monitoring and create spontaneous confabulations

psychological or behavioral processes. Neural *coding* is inferred in practice by measuring brain *activity correlated to a pleasure*, using techniques such as PET, fMRI and MEG neuroimaging in humans, or electrophysiological or neurochemical activation measures in animals presented with a rewarding stimulus. Causation is generally inferred on the basis of a *change* in pleasure as a *consequence of a brain manipulation* such as lesion or stimulation.

As a general rule, we suggest that brains operate by the principle of 'many more codes than causes' for pleasure. In part, the greater number of hedonic coding sites results from the tendency of signals to spread beyond their source, as well as from the massive need for brain systems to translate pleasure signals into many other psychological functions, such as learning and memory, cognitive representations, decisions, action, and consciousness.

Code-but-not-cause systems might nonetheless be reliable indicators that a pleasant event is occurring, because they must take pleasure signals as inputs to achieve other component processes in reward and related. We distinguish here between the cognitive representations and memories of reward (reward learning) and the motivational value appraisals or decisions (reward wanting). For example, parts of the prefrontal cortex regions sensitively code reward and hedonic impact, as described below. Yet damage to ventromedial region of prefrontal cortex may impair the cognitive use of emotional reactions without necessarily impairing the capacity to experience the hedonic impact of those emotional reactions (Bechara et al. 1997; Damasio 2004; Kringelbach 2005). The difference between coding and causing poses challenges to human affective neuroscience studies, where lesions, stimulations or other causal tools are rarely available.

7.4 Cortical Cognition and Pleasure

In humans, evidence suggests that pleasure encoding may reach an apex of cortical localization in a subregion of orbitofrontal cortex: this hedonic-coding site is placed in the mid-anterior and roughly mid-lateral zone of the orbitofrontal region. Here neuroimaging activity in people particularly correlates strongly to subjective pleasantness ratings of food varieties—and to other pleasures such as sexual orgasms, drugs, chocolate, and music (Geogiadis and Kortekaas 2010; Kringelbach and Berridge 2010a; Leknes and Tracey 2010; Veldhuizen et al. 2010; Vuust and Kringelbach 2010). Most importantly, activity in this special mid-anterior zone of orbitofrontal cortex tracks changes in subjective pleasure, such as a decline in palatability when the reward value of one food was reduced by eating it to satiety (while remaining high to another food). The mid-anterior subregion of orbitofrontal cortex is thus a prime candidate for the coding of subjective experience of pleasure (Kringelbach 2005).

Another potential coding site for positive hedonics in orbitofrontal cortex is along its medial edge that has activity related to the positive and negative valence of affective events, contrasted to lateral portions that have been suggested to code unpleasant events (although lateral activity may reflect a signal to escape the situation, rather than displeasure per se) (Kringelbach 2010; Kringelbach and Rolls 2004). This medial–lateral hedonic gradient in orbitofrontal cortex interacts with an abstraction-concreteness gradient in the posterior-anterior dimension, so that more complex or abstract reinforcers (such as monetary gain and loss) are represented more anteriorly in the orbitofrontal cortex than less complex sensory rewards (such as taste). The medial region that codes pleasant sensations does not, however, appear to change its activity with reinforcer devaluation as effectively as the mid-anterior subregion that best codes hedonics, and so the medial region may not reflect the full dynamics of pleasure.

A malfunction of these hedonic mechanisms in the orbitofrontal cortex could explain the profound changes in eating habits (escalating desire for sweet food coupled with reduced satiety) that are often followed by enormous weight gain in patients with frontotemporal dementia. This progressive neurodegenerative disorder is associated with major and pervasive behavioural changes in personality and social conduct resembling those produced by orbitofrontal lesions (although it should be noted that more focal lesions to the orbitofrontal cortex have not to date been associated with obesity) (Rahman et al. 1999). It has become clear recently that the orbitofrontal cortex also has an important role in emotional disorders such as depression and addiction (Kringelbach 2005).

Beyond orbitofrontal cortex, other cortical regions implicated in coding for pleasant stimuli include parts of the mid-insular (Craig 2009) and anterior cingulate cortices. As yet, however, it is not as clear as for the orbitofrontal cortex whether those regions specifically code pleasure or only emotion more generally (Feldman et al. 2006). A related suggestion has emerged that the frontal left hemisphere plays a special lateralized role in positive affect more than the right hemisphere (Davidson 2004). Most specifically related to well-being, resting EEG activity in left prefrontal cortex has been reported to be higher in individuals with greater eudaimonic and hedonic well-being (Urry et al. 2004). How to reconcile left-positive findings with many other findings of bilateral activity in orbitofrontal and related cortical regions during hedonic processing remains an ongoing puzzle.

Cortical Causation of Human Pleasure?

Despite the evidence above for hedonic coding, however, it still remains unknown if even the mid-anterior pleasure-coding site of orbitofrontal cortex actually *causes* a positive pleasure state. It would be of considerable interest to investigate whether any of these sub-regions of the orbitofrontal cortex are necessary or sufficient causes of pleasure, or alternatively whether their role is restricted to cognitive elaboration of value, and translation of hedonic affect into goal-directed plans.

The proposed link to subjective hedonic processing might make the orbitofrontal cortex an important gateway for neuroscientific analyses of human subjective conscious experience. Some have even suggested that the orbitofrontal and anterior cingulate cortices could be viewed as part of a global workspace for access to consciousness with the specific role of evaluating the affective valence of stimuli (Dehaene et al. 1998; Kringelbach and Berridge 2010a). In this context, it is interesting that the medial parts of the orbitofrontal are part of a proposed network for the baseline activity of the human brain at rest (Gusnard et al. 2001), as this would place the orbitofrontal cortex as a key node in the network subserving consciousness. This could potentially explain why all our subjective experiences have an emotional tone and perhaps even why we have conscious pleasure.

One way of investigating this causation question would be to ask whether the orbitofrontal cortex is actually required for normal pleasure reactions or conscious feelings. Only scattered data are available, primarily from historical and case study sources. Prefrontal lobotomies were performed on thousands of human patients in the 1950s, and may provide some insights. If orbitofrontal or other prefrontal areas are necessary for basic 'liking' reactions, these lobotomy patients should no longer have been able to feel pleasure. Yet perhaps surprisingly from this perspective, prefrontal lobotomy may not produce a catastrophic loss of pleasure feelings as far as one can tell from the available literature. Although many subtle emotional deficits occur in how patients describe or act upon their emotions after damage to prefrontal cortex the capacity for basic 'liking' reactions appeared to remain intact. Lobotomy patients were by no means oblivious to the pleasures of food, sex or other rewards.

Modern analyses of more focal prefrontal lesions report deficits in cognitive-emotional processing of decisions of human patients, similarly do not indicate a total loss of the capacity for pleasures (Bechara et al. 2000; Damasio 1999; Damasio 2004; Hornak et al. 2003). Decisions are often profoundly imbalanced in such patients but positive hedonia does not seem abolished by medial prefrontal or orbitofrontal cortex lesions.

Such considerations suggest that orbitofrontal cortex might be more important to translating hedonic information into cognitive representations and decisions than to generating a core 'liking' reaction to pleasant events (Burke et al. 2010; Dickinson and Balleine 2010).

Such evidence leads us to suggest that that the human prefrontal cortex might not actually be needed to cause pleasure, or at least not all pleasures. It is possible that the main role of the prefrontal cortex in pleasure is to act as the interface of higher order processing such as consciousness and attention to the non-conscious pleasure generators in primarily sub-cortical regions.

At its extreme, this position might view hedonic reactions as arising from subcortical structures even when the subcortical brain is on its own and unable to interact with neocortex. Some further evidence from humans, as well as much from animals, supports a subcortical emphasis for pleasure generation. For example, Shewmon et al. described several hydranencephalic cases, including a 6-year old boy with congenital "absence of cerebral tissue rostral to the thalamus, except for small mesial temporal-lobe remnants" (Shewmon et al. 1999, p. 364) and a tissue-lined cyst, who nevertheless "smiled when spoken to and giggled when played with. These human interactions were much more intense than, and qualitatively different from, his positive reactions to favorite toys and music" (Shewmon et al. 1999, p. 366). Similarly, Merker suggested that hydranencephalic children "express pleasure by smiling and laughter, and aversion by "fussing," arching of the back and crying (in many gradations), their faces being animated by these emotional states. A familiar adult can employ this responsiveness to build up play sequences predictably progressing from smiling, through giggling, to laughter and great excitement on the part of the child." (Merker 2007, p. 79).

Such cases of emotional reaction without (much) cortex raise fascinating questions for future consideration about the relative roles of cortical regions versus subcortical structures in human pleasures. However, no matter what conclusion is reached regarding pleasure generation, there seems general consensus that neocortex is crucial to link affect with complex cognition and introspection about hedonic states.

Controversial Subcortical Pleasure Generators

Several other particular limbic substrates, even subcortical ones, which were once thought to cause pleasure have now turned out not to do so after all. These include the mesolimbic dopamine system and many so-called pleasure electrodes in related brain substrates.

Mesolimbic dopamine was long regarded as a pleasure neurotransmitter, but now is increasingly thought by many neuroscientists to fail to live up to its pleasure label. Instead, dopamine is currently thought by many to facilitate some psychological valuation process besides either learning or pleasure 'liking'. Suggestions have included motivational incentive salience, arousal, motivation, and memory consolidation. We think it safe to sum up by saying that the consensus among affective neuroscientists today is that brain mesolimbic dopamine is not, after all, primarily a pleasure neurotransmitter.

Similarly, 'pleasure electrodes' in the brain for 50 years were supposed to tap directly into brain pleasure circuits (Olds 1956). However, we believe that many so-called 'pleasure electrodes' may actually have failed to truly cause significant pleasure at all (Kringelbach and Berridge 2012). Instead we suggest most electrodes more exclusively activated only the 'wanting' component of reward (similar to mesolimbic dopamine stimulation; which indeed is typically activated by such electrodes). Such electrode activations may be sought out, or may stimulate seeking of external rewards (food, sex, gambling, shopping, etc.), yet need not be pleasant themselves.

7.5 Towards a Balanced Brain

It is interesting to note that all brain structures discussed above or being targeted for brain-based treatments of pathological mood disorders today either have close links with the hedonic network we have considered (e.g., orbitofrontal cortex, nucleus accumbens and ventral pallidum, etc.) or belong to what has been termed the brain's default network which changes over early development (e.g., additional regions of prefrontal cortex, or of cingulate cortex, temporal cortex, and parietal cortex) (Fair et al. 2008; Fransson et al. 2007) (Fig. 7.4).

Mention of the default network brings us back to the topic of eudaimonic happiness, and to potential interactions of hedonic brain circuits with circuits that assess meaningful relationships of self to social others. The default network is a steady state circuit of the brain which becomes perturbed during cognitive tasks (Gusnard et al. 2001). Most pertinent here is an emerging literature that has proposed the default network to carry representations of self (Lou et al. 1999), internal modes of cognition (Buckner et al. 2008), and perhaps even states of consciousness (Laureys et al. 2004). Such functions might well be important to higher pleasures as well as meaningful aspects of happiness.

Although highly speculative, we wonder whether the default network might deserve further consideration for a role in connecting eudaimonic and hedonic happiness. At least, key regions of the frontal default network overlap with the hedonic network discussed above, such as the anterior cingulate and orbitofrontal cortices, and have a relatively high density of opiate receptors. Eudaimonic well being may be correlated with activity in the anterior cingulate and in left prefrontal cortex, perhaps though the ability to suppress negative emotions (Urry et al. 2004; Urry et al. 2006; van Reekum et al. 2007). Activity changes in the frontal default network, such as in the subgenual cingulate and orbitofrontal cortices, correlate to pathological changes in subjective hedonic experience, such as in depressed patients (Davidson et al. 2002).

Pathological self-representations by the frontal default network could also provide a potential link between hedonic distortions of happiness that are accompanied by eudaimonic dissatisfaction, such as in cognitive rumination of depression. Conversely, mindfulness-based cognitive therapy for depression, which aims to disengage from dysphoria-activated depressogenic thinking might conceivably recruit default network circuitry to help mediate improvement in happiness via a linkage to hedonic circuitry.

Beyond the default network are other cortical networks in which activations may correspond with evaluations of self, others, and meaningful themes related to life satisfaction (Heller et al. 2009; Schacter et al. 2007). These include dorsolateral prefrontal, and other parietal and temporal areas of cortex and related networks. In short, the default network and networks whose activation encodes evaluations of self and life meaning stand among the brain candidates for a substrate that might mediate eudaimonia appraisals. How these networks might embody eudaimonia components, and link evaluations of life meaningfulness and satisfaction with pleasurable states of hedonia, remains a major challenge to psychological neuroscience for the future.

7.6 Conclusions

As shown in this review, the main role of pleasure is to help initiate, sustain or terminate the phases involved in pleasure cycles of reward. Pleasure can thus be said to play a crucial role in guiding the survival-related decision-making involved in optimizing resource allocation of brain resources. From this perspective *optimization* rather than *maximization* of pleasure processing is the most sensible strategy since this leads to the most optimal brain resource allocation.

It is not straightforward, however, to bring this balancing view of hedonia a step further to understand the relation of sensory pleasure to the more enduring hedonia of well-being, the interaction between hedonia (pleasure or positive affect) and eudaimonia (cognitive appraisals of meaning and life satisfaction), within underlying brain systems, and the nature of their subjective experience.

While some progress has been made in understanding brain hedonics, it is important not to over-interpret the findings. In particular we have still not made substantial progress towards understanding the functional neuroscience specifically of well-being or happiness. We have merely aimed to sketch out the beginnings of a hedonic approach.

Further, when all is done, one may still question our entire effort, based as it is largely on evidence from sensory pleasures. Some will demur that pleasure, our chief focus here, is irrelevant after all to true happiness. For many, this view might be well expressed by the words of John Stuart Mill, "It is better to be a human being dissatisfied than a pig satisfied; better to be Socrates dissatisfied than a fool satisfied" (Mill et al. 1998, p. 57). By the view expressed in this quotation, a life filled with the most intense pleasures of pigs or fools would never be enough for happiness. That is because true happiness hinges on a superior kind of psychological or eudaimonic richness that is unique to the enlightened, though hedonically dissatisfied, Socrates. But note that Mill, however, seemed to say elsewhere that hedonic pleasure was important to happiness too.

At the opposite extreme, Sigmund Freud seemed to take a purely hedonic view of happiness, more likely to favor our endeavor. Freud wrote, in response to his own question about what people demand of life and wish to achieve in it, the reply "The answer to this can hardly be in doubt. They strive after happiness; they want to become happy and to remain so. This endeavor has two sides, a positive and a negative aim. It aims, on the one hand, at an absence of pain and displeasure, and, on the other, at the experiencing of strong feelings of pleasure" (Freud 1930, p. 76). Freud's answer equates hedonic pleasure with happiness. According to this view, the more pleasure you have (while avoiding displeasure), the happier you are. Modern psychologists tend to fall in between these poles. Yet relatively few today would deny that hedonic pleasure is at least relevant to a final state of well-being.

We do not pretend to see deeper into the nature of happiness than such major figures of earlier times, but simply point again to the empirical convergence of hedonic and eudaimonic features together in most people who are actually happy. And we note in conclusion, that so far as positive affect contributes to happiness, then at least some progress has been made in understanding the neurobiology of pleasure in ways that might be relevant.

In finishing, we can imagine several possibilities to relate happiness to particular hedonic psychological processes discussed above. Thus, one way to conceive of hedonic happiness is as 'liking' without 'wanting'. That is, a state of pleasure without disruptive desires, a state of contentment (Kringelbach and Berridge 2009). Another possibility is that moderate 'wanting', matched to positive 'liking', facilitates engagement with the world. A little incentive salience may add zest to the perception of life and perhaps even promote the construction of meaning, just as in some patients therapeutic deep brain stimulation may help lift the veil of depression by making life events more appealing. However, too much 'wanting' can readily spiral into maladaptive patterns such as addiction, and is a direct route to great unhappiness. Finally, all might agree that happiness springs not from any single component but from the interplay of higher pleasures, positive appraisals of life meaning and social connectedness, all combined and merged by interaction between the brain's default networks and pleasure networks. Achieving the right hedonic balance in such ways may be crucial to keep one not just moving forward through life—but even to achieve high state of well-being.

Acknowledgments Our research is supported by grants from the TrygFonden Charitable Foundation, Braveheart Charity, and Novo Nordisk Foundation to MLK and from the NIH (MH63644 and DA015188) to KCB. This chapter is an abridged version of a previously published article (Berridge and Kringelbach 2011).

References

Aldridge JW, Berridge KC, Herman M, Zimmer L (1993) Neuronal coding of serial order: syntax of grooming in the neostriatum. Psychol Sci 4:391–395

Aristotle (350 B. C. [2009]) The Nicomachean ethics. Oxford University Press, Oxford

Bechara A, Damasio H, Tranel D, Damasio AR (1997) Deciding advantageously before knowing the advantageous strategy. Science 275:1293–1295

Bechara A, Damasio H, Damasio AR (2000) Emotion, decision making and the orbitofrontal cortex. Cereb Cortex 10:295–307

Berridge KC, Fentress JC (1986) Contextual control of trigeminal sensorimotor function. J Neurosci 6:325–330

Berridge KC, Kringelbach ML (2008) Affective neuroscience of pleasure: reward in humans and animals. Psychopharmacology 199:457–480

Berridge KC, Kringelbach ML (2011) Building a neuroscience of pleasure and well-being. Psychol Well-Being 1:1–3

Berridge KC, Robinson TE, Aldridge JW (2009) Dissecting components of reward: 'liking', 'wanting', and learning. Curr Opin Pharmacol 9:65–73

Berridge KC, Ho CY, Richard JM, DiFeliceantonio AG (2010) The tempted brain eats: pleasure and desire circuits in obesity and eating disorders. Brain Res 1350:43–64

Bruno M-A, Bernheim JL, Ledoux D, Pellas F, Demertzi A, Laureys S (2011) A survey on self-assessed well-being in a cohort of chronic locked-in syndrome patients: Happy majority, miserable minority. Brit Med J Open. 1(1):e000039

Buckner RL, Andrews-Hanna JR, Schacter DL (2008) The brain's default network: anatomy, function, and relevance to disease. Ann N Y Acad Sci 1124:1–38

Burke KA, Miller D, Schoenbaum G (2010) Conditioned reinforcement and the specialized role of corticolimbic circuits in the pursuit of happiness and other more specific rewards. In: Kringelbach ML, Berridge KC (eds) Pleasures of the brain. Oxford University Press, Oxford, pp 50–61

Cabanac M (2010) The dialectics of pleasure. In: Kringelbach ML, Berridge KC (eds) Pleasures of the brain. Oxford University Press, Oxford, pp 113–124

Craig AD (2009) How do you feel–now? The anterior insula and human awareness. Nat Rev Neurosci 10:59–70

Cromwell HC, Berridge KC (1993) Where does damage lead to enhanced food aversion: the ventral pallidum/substantia innominata or lateral hypothalamus? Brain Res 624:1–10

Damasio AR (1999) The feeling of what happens: body and emotion in the making of consciousness. Harcourt Brace, New York

Damasio AR (2004) Emotions and feelings: a neurobiological perspective. In: Manstead ASR, Frijda N, Fischer A (eds) Feelings and emotions: the Amsterdam symposium. Cambridge University Press, Cambridge, pp 49–57

Darwin C (1872) The expression of the emotions in man and animals. J. Murray, London
Davidson RJ (2004) Well-being and affective style: neural substrates and biobehavioural correlates. Philos Trans R Soc B: Biol Sci 359:1395–1411
Davidson RJ, Lewis DA, Alloy LB, Amaral DG, Bush G, Cohen JD et al (2002) Neural and behavioral substrates of mood and mood regulation. Biol Psychiatry 52:478–502
Dehaene S, Kerszberg M, Changeux JP (1998) A neuronal model of a global workspace in effortful cognitive tasks. Proc Natl Acad Sci U S A 95:14529–14534
Dickinson A, Balleine B (2010) Hedonics: the cognitive-motivational interface. In: Kringelbach ML, Berridge KC (eds) Pleasures of the brain. Oxford University Press, Oxford, pp 74–84
Diener E, Lucas RE, Scollon CN (2006) Beyond the hedonic treadmill: revising the adaptation theory of well-being. Am Psychol 61:305–314
Diener E, Kesebir P, Lucas R (2008) Benefits of accounts of well-being—for societies and for psychological science. Appl Psychol 57:37–53
Everitt BJ, Belin D, Economidou D, Pelloux Y, Dalley JW, Robbins TW (2008) Neural mechanisms underlying the vulnerability to develop compulsive drug-seeking habits and addiction. Philos Trans R Soc B: Biol Sci 363:3125–3135
Fair DA, Cohen AL, Dosenbach NU, Church JA, Miezin FM, Barch DM et al (2008) The maturing architecture of the brain's default network. Proc Natl Acad Sci U S A 105:4028–4032
Feldman BL, Wager TD (2006) The structure of emotion: evidence from neuroimaging studies. Curr Dir Psychol Sci 15:79–83
Fransson P, Skiold B, Horsch S, Nordell A, Blennow M, Lagercrantz H, Aden U (2007) Resting-state networks in the infant brain. Proc Natl Acad Sci U S A 104:15531–15536
Freud S (1930) Civilization and its discontents. Norton & Co., New York
Frijda N (2010) On the nature and function of pleasure. In: Kringelbach ML, Berridge KC (eds) Pleasures of the brain. Oxford University Press, Oxford, pp 99–112
Friston K, Kiebel S (2009) Predictive coding under the free-energy principle. Philos Trans R Soc B: Biol Sci 364:1211–1221
Frith CD, Frith U (1999) Interacting minds–a biological basis. Science 286:1692–1695
Geogiadis JR, Kortekaas R (2010) The sweetest taboo: functional neurobiology of human sexuality in relation to pleasure. In: Kringelbach ML, Berridge KC (eds) Pleasures of the brain. Oxford University Press, Oxford, pp 178–291
Georgiadis JR, Kringelbach ML (2012) The human sexual response cycle: brain imaging evidence linking sex to other pleasures. Prog Neurobiol 98(1):49–81
Gilbert DT (2006) Stumbling on happiness. Alfred A. Knopf, New York
Grabenhorst F, Rolls ET (2011) Value, pleasure and choice in the ventral prefrontal cortex. Trends Cogn Sci 15:56–67
Grill HJ, Norgren R (1978) The taste reactivity test. II. Mimetic responses to gustatory stimuli in chronic thalamic and chronic decerebrate rats. Brain Res 143:281–297
Gusnard DA, Raichle ME, Raichle ME (2001) Searching for a baseline: functional imaging and the resting human brain. Nat Rev Neurosci 2:685–694
Harris S, Kaplan JT, Curiel A, Bookheimer SY, Iacoboni M, Cohen MS (2009) The neural correlates of religious and nonreligious relief. PLoS ONE 4(10):e0007272
Heller AS, Johnstone T, Shackman AJ, Light SN, Peterson MJ, Kolden GG et al (2009) Reduced capacity to sustain positive emotion in major depression reflects diminished maintenance of fronto-striatal brain activation. Proc Natl Acad Sci U S A 106:22445–22450
Hornak J, Bramham J, Rolls ET, Morris RG, O'Doherty J, Bullock PR, Polkey CE (2003) Changes in emotion after circumscribed surgical lesions of the orbitofrontal and cingulate cortices. Brain 126:1691–1712
Kahneman D (1999a) Assessments of individual well-being: a bottom-up approach. In: Kahneman D, Diener E, Schwartz N (eds) Well-being: the foundations of hedonic psychology. Russel Sage Foundation, New York
Kahneman D (1999b) Objective happiness. In: Kahneman D, Diener E, Schwartz N (eds) Well-being: the foundations of hedonic psychology. Russel Sage Foundation, New York, pp 3–25

Kelley AE, Berridge KC (2002) The neuroscience of natural rewards: relevance to addictive drugs. J Neurosci 22:3306–3311
Koob GF, Volkow ND (2010) Neurocircuitry of addiction. Neuropsychopharmacology 35:217–238
Kringelbach ML (2004) Food for thought: hedonic experience beyond homeostasis in the human brain. Neuroscience 126:807–819
Kringelbach ML (2005) The human orbitofrontal cortex: linking reward to hedonic experience. Nat Rev Neurosci 6:691–702
Kringelbach ML (2009) The pleasure center: trust your animal instincts. Oxford University Press, New York
Kringelbach ML (2010) The hedonic brain: a functional neuroanatomy of human pleasure. In: Kringelbach ML, Berridge KC (eds) Pleasures of the brain. Oxford University Press, Oxford, pp 202–221
Kringelbach ML, Berridge KC (2009) Towards a functional neuroanatomy of pleasure and happiness. Trends Cogn Sci 13:479–487
Kringelbach ML, Berridge KC (2010a) The functional neuroanatomy of pleasure and happiness. Discov Med 9:579–587
Kringelbach ML, Berridge KC (eds) (2010b) Pleasures of the brain. Oxford University Press, Oxford
Kringelbach ML, Berridge KC (2012) A joyful mind. Scientific American, USA
Kringelbach ML, Rolls ET (2004) The functional neuroanatomy of the human orbitofrontal cortex: evidence from neuroimaging and neuropsychology. Prog Neurobiol 72:341–372
Kuppens P, Realo A, Diener E (2008) The role of positive and negative emotions in life satisfaction judgment across nations. J Pers Soc Psychol 95:66–75
Laureys S, Perrin F, Faymonville ME, Schnakers C, Boly M, Bartsch V et al (2004) Cerebral processing in the minimally conscious state. Neurology 63:916–918
Leknes S, Tracey I (2010) Pleasure and pain: masters of mankind. In: Kringelbach ML, Berridge KC (eds) Pleasures of the brain. Oxford University Press, Oxford, pp 320–335
Lou HC, Kjaer TW, Friberg L, Wildschiodtz G, Holm S, Nowak M (1999) A 15O–H2O PET study of meditation and the resting state of normal consciousness. Hum Brain Mapp 7:98–105
Mahler SV, Smith KS, Berridge KC (2007) Endocannabinoid hedonic hotspot for sensory pleasure: Anandamide in nucleus accumbens shell enhances 'liking' of a sweet reward. Neuropsychopharmacology 32:2267–2278
Merker B (2007) Consciousness without a cerebral cortex: a challenge for neuroscience and medicine. Behav Brain Sci 30:63–81
Mill JS, Crisp R, NetLibrary Inc. (1998) Utilitarianism. In: Oxford philosophical texts, Oxford University Press, Oxford, pp vii, 157
Miller J, Vorel S, Tranguch A, Kenny E, Mazzoni P, van Gorp W, Kleber H (2006) Anhedonia after a selective bilateral lesion of the globus pallidus. Am J Psychiatry 163:786–788
Nesse RM (2002) Evolutionary biology: a basic science for psychiatry. World Psychiatry 1:7–9
Olds J (1956) Pleasure centers in the brain. Sci Am 195:105–116
Panksepp J (1998) Affective neuroscience: the foundations of human and animal emotions. Oxford University Press, Oxford
Pecina S (2008) Opioid reward 'liking' and 'wanting' in the nucleus accumbens. Physiol Behav 94:675–680
Peciña S, Berridge KC (2005) Hedonic hot spot in nucleus accumbens shell: where do mu-opioids cause increased hedonic impact of sweetness? J Neurosci 25:11777–11786
Peciña S, Smith KS (2010) Hedonic and motivational roles of opioids in food reward: Implications for overeating disorders. Pharmacol Biochem Behav 97:34–46
Peciña S, Smith KS, Berridge KC (2006) Hedonic hot spots in the brain. Neuroscientist 12:500–511
Rahman S, Robbins TW, Sahakian BJ (1999) Comparative cognitive neuropsychological studies of frontal lobe function: implications for therapeutic strategies in frontal variant frontotemporal dementia. Dement Geriatr Cogn Disord 10(Supp. 1):15–28

Ryle G (1954) Pleasure. Proc Aristotelian Soc 28:135–146

Salimpoor VN, Benovoy M, Larcher K, Dagher A, Zatorre RJ (2011) Anatomically distinct dopamine release during anticipation and experience of peak emotion to music. Nat Neurosci 14:257–262

Schacter DL, Addis DR, Buckner RL (2007) Remembering the past to imagine the future: the prospective brain. Nat Rev Neurosci 8:657–661

Seligman MEP, Steen TA, Park N, Peterson C (2005) Positive psychology progress—empirical validation of interventions. Am Psychol 60:410–421

Shewmon DA, Holmes GL, Byrne PA (1999) Consciousness in congenitally decorticate children: Developmental vegetative state as self-fulfilling prophecy. Dev Med Child Neurol 41:364–374

Skov M (2010) The pleasures of art. In: Kringelbach ML, Berridge KC (eds) Pleasures of the brain. Oxford University Press, Oxford, pp 270–286

Smith KS, Berridge KC (2005) The ventral pallidum and hedonic reward: neurochemical maps of sucrose "liking" and food intake. J Neurosci 25:8637–8649

Smith KS, Berridge KC (2007) Opioid limbic circuit for reward: Interaction between hedonic hotspots of nucleus accumbens and ventral pallidum. J Neurosci 27:1594–1605

Smith KS, Mahler SV, Pecina S, Berridge KC (2010) Hedonic hotspots: generating sensory pleasure in the brain. In: Kringelbach ML, Berridge KC (eds) Pleasures of the brain. Oxford University Press, Oxford, pp 27–49

Steiner JE, Glaser D, Hawilo ME, Berridge KC (2001) Comparative expression of hedonic impact: Affective reactions to taste by human infants and other primates. Neurosci Biobehav Rev 25:53–74

Urry HL, Nitschke JB, Dolski I, Jackson DC, Dalton KM, Mueller CJ et al (2004) Making a life worth living: neural correlates of well-being. Psychol Sci 15:367–372

Urry HL, van Reekum CM, Johnstone T, Kalin NH, Thurow ME, Schaefer HS et al (2006) Amygdala and ventromedial prefrontal cortex are inversely coupled during regulation of negative affect and predict the diurnal pattern of cortisol secretion among older adults. J Neurosci 26:4415–4425

van Reekum CM, Urry HL, Johnstone T, Thurow ME, Frye CJ, Jackson CA et al (2007) Individual differences in amygdala and ventromedial prefrontal cortex activity are associated with evaluation speed and psychological well-being. J Cogn Neurosci 19:237–248

Veldhuizen MG, Rudenga KJ, Small D (2010) The pleasure of taste flavor and food. In: Kringelbach ML, Berridge KC (eds) Pleasures of the brain. Oxford University Press, Oxford, pp 146–168

Vuust P, Kringelbach ML (2010) The pleasure of music. In: Kringelbach ML, Berridge KC (eds) Pleasures of the brain. Oxford University Press, Oxford, pp 255–269

Winkielman P, Berridge KC, Wilbarger JL (2005) Unconscious affective reactions to masked happy versus angry faces influence consumption behavior and judgments of value. Pers Soc Psychol Bull 31:121–135

Zhang J, Berridge KC, Tindell AJ, Smith KS, Aldridge JW (2009) A neural computational model of incentive salience. PLoS Comput Biol 5(7):e1000437

Chapter 8
From Treating Mental Dysfunction to Neuroenhancement

Michael Koch

8.1 Introduction

Happiness is a complex psychosocial construct in humans, involving short moments of luck or serendipity, longer lasting states of joy, pleasure, and the reflection on leading a healthy, rich life in the future (see Chap. 1). Several *mental* processes contribute to these different aspects of happiness, for example the processing of reward-related stimuli and contexts, learning and memory, or the executive functions enabling intelligent planning, behavioral control and social decision making (Loewenstein et al. 2008, Platt and Huettel 2008). This is bitterly experienced by people suffering from neurological or psychiatric disorders that impair the functioning of the brain. Almost all of these pathological conditions—not only the canonical mood disorders, such as depression and anxiety—lead to a severe reduction in the experience of happiness and have motivated pharmacological research.

8.2 The Pharmacology of Mental Dysfunctions and Happiness[1]

Progress in psychopharmacology over the past thirty years has led to increased possibilities to treat various forms of mental dysfunction as well as to enhance cognitive performance and happiness in general. Many of these new drugs show

[1] This chapter deals with pharmacological means of neuro-enhancement only. I will not describe the possibility of cyborgs or the use of brain or spinal cord implants ("Orgasmatron") and neuroprosthetics that may be used to electronically mimic or support the functions of the central nervous system.

M. Koch (✉)
Brain Research Institute, Department of Neuropharmacology, Center for Cognitive Sciences COGNIUM, University of Bremen, Bremen, Germany
e-mail: michael.koch@uni-bremen.de

H. Brockmann and J. Delhey (eds.), *Human Happiness and the Pursuit of Maximization*, Happiness Studies Book Series, DOI: 10.1007/978-94-007-6609-9_8,

relatively few side effects and are well tolerated by patients. Neurological and psychiatric disorders are usually complex diseases with variable etiologies, symptoms and neuropathologies. Therefore, there are a variety of compounds that are being used for the treatment of these conditions. Many mental states that contribute to happiness are potential targets of drugs, including "basic drives" such as appetite, libido and sleep, as well as "higher cognitive functions" such as attention, learning, memory, executive functions and social behavior.

All mental functions depend on communication within networks of nerve cells (Kandel 1991). The information transfer between nerve cells in the brain usually occurs through the action of chemical messengers, or neurotransmitters. The main aim of neuro- and psychopharmacology is to influence this chemical communication in a rather selective way by drugs in order to mitigate deficits that occur in disease states. Chemical communication in the nervous system is mediated by the so-called "classical" transmitters, such as acetylcholine, glutamate or dopamine. In addition, there are many different neuropeptides (e.g., oxytocin or the endogeneous opioids, such as β-endorphin) and various neuromodulators (such as steroid hormones, the endogenous cannabinoids or the ubiquitous cellular nucleoside adenosine) that either modulate the action of classical transmitters or have actions on their own. Furthermore, several other types of molecules are necessary to maintain the function of nerve cells, e.g., extracellular matrix proteins, neurotrophic factors and cell adhesion molecules. This provides a broad range of targets that can be hit by drugs (Cooper et al. 2003; Davis et al. 2002; von Bohlen und Halbach and Dermietzel 2002).

Neurotransmitters, neuropeptides and neuromodulators are either organized in neuroanatomically specific systems (e.g., the monoamine transmitters dopamine, noradrenaline or serotonin) or they occur more or less ubiquitously in the brain (e.g., glutamate, the main excitatory transmitter, and γ-amino butyric acid, the main inhibitory transmitter).

Chemical communication in the nervous system usually occurs at specialized contact points between neurons, the so-called *synapses*, and is comprised of a series of molecular and physiological events: Neurotransmitter molecules are synthesized by specific enzymes in the presynaptic ending of the neuron and packed into small vesicles by transporter proteins. When the neuron is excited, it sends an action potential (depolarization of the cell membrane) along its axon. At the presynaptic ending of this axon, the action potential leads to an influx of calcium into the neuron, which triggers fusion of these vesicles with the presynaptic membrane (by activation of a specific fusion complex composed of proteins) and release of the neurotransmitter into the synaptic cleft.

The transmitter molecules now diffuse around their site of release and thereby reach the membrane of another (postsynaptic) neuron, where they bind to specific receptor proteins. The binding of a transmitter to a receptor triggers a cascade of intracellular events that eventually alter the activity of the postsynaptic neuron by changing ion fluxes in and out of the cell. Transmitters can depolarize or hyperpolarize a cell, i.e. their action can be either excitatory or inhibitory. However, transmitter molecules also diffuse back to the presynaptic neuron, where they can bind to so-called auto-receptors that regulate their release, or where they are taken up into the presynaptic neuron by transporter molecules. There are also enzymes in the presynaptic terminal and in the synaptic cleft that rapidly degrade the neurotransmitters (Cooper et al. 2003). Each of these key players in synaptic communication between nerve cells can be the target of psychopharmacological agents.

8.3 Curing Chemicals and Pleasing Pills

The fusion proteins essential for transmitter exocytosis can be inactivated by several toxins (e.g., the botulinum toxins, aka "Botox"), so that no transmitter can be released. Agonists or antagonists of transmitter receptors can either mimic or block, respectively, the cellular effect of the transmitter. For example, pramipexole is a dopamine receptor agonist that can be used for the treatment of Parkinson's disease, where there is a lack of dopamine in some parts of the brain. Haloperidol is a dopamine receptor antagonist that can be used as an antipsychotic compound because it blocks the excessive activity of dopamine that occurs in some psychoses. The interaction of a transmitter with its receptors can also be modulated by certain compounds. For example the benzodiazepines (e.g., diazepam aka valium) bind to a certain receptor for the neurotransmitter gamma-aminobutyric acid (GABA) and thereby enhance the efficacy of GABA. Benzodiazepines are sedatives that are widely used for the symptomatic treatment of fear- and anxiety-related disorders and during alcohol withdrawal.

Inhibitors of transmitter-metabolizing enzymes have been used to enhance the action of the transmitter due to its reduced rate of metabolism. For example, inhibitors of the enzyme acetylcholinesterase prevent or reduce the metabolism of the transmitter acetylcholine and thereby enhance its effects. Acetylcholine is important for the maintenance of several cognitive functions, for example learning, memory and attention. Donepezil is a reversible inhibitor of acetylcholinesterase and has been used as a cognitive enhancer in dementias - including Alzheimer's disease - that are characterized by a loss of cholinergic activity. Inhibitors of the enzyme monoamine oxidase (MAO), which metabolizes transmitters such as dopamine, noradrenaline and serotonin, have been used for the treatment of mood disorders and Parkinson's disease.

The transport back into the cell of transmitter molecules that have been released into the synaptic cleft by highly selective transporter proteins is a very efficient and rapid way to terminate the action of the transmitter. Therefore, in the case of a disease where the symptoms are caused by a lack of a certain transmitter, it is useful to inhibit the reuptake transporter for this transmitter. This mode of action has been utilized by the so-called tricyclic antidepressant drugs, and also more recently by selective serotonin or noradrenaline reuptake inhibitors (SSRI and SNRI, respectively) that increase the amount of serotonin or noradrenaline in the synaptic cleft, thereby ameliorating depressive symptoms.

Another drug that inhibits monoamine transporters is methylphenidate (Ritalin), used to treat attention deficit—hyperactivity disorder (ADHD). This drug has some specificity for blocking the dopamine transporter, a property Ritalin shares with cocaine and amphetamine, and leads to a more or less selective increase of dopamine in the frontal cortex with beneficial effects on the symptoms of ADHD. In healthy people, Ritalin is known to exert activating and rewarding effects. The worldwide consumption rates of Ritalin have increased dramatically in the past years. Iceland, the US, Canada and the UK have the highest consumption rates, with about 5 daily doses per 1,000 inhabitants. The nonprescription use of this drug has increased from below 0.5 % in 1995 up to more than 5 % in college students about twenty years later (McCabe et al. 2005; Singh 2008; Singh and Kelleher 2010; Smith and Farah 2011).

Psychopharmacological research has led to striking refinements of the physiological effects of drugs which capitalize on their relatively complex pharmacology. For example, while the classical first-generation neuroleptic drugs like chlorpromazine and haloperidol mainly acted as dopamine receptor blockers (Creese et al. 1976), modern, so-called atypical antipsychotic compounds (e.g., clozapine, quetiapin, risperidon or aripiprazole) act as antagonists or partial agonists at various transmitter receptors with different patterns of affinities, which enables relatively selective treatment of the various different symptoms of schizophrenia (Koch 2007). Another example of research attempts to refine a drug's effects is the design of anxiolytic drugs without the muscle relaxant, mnemonic, addictive and sedative side effects of the benzodiazepines (McKernan et al. 2000).

8.4 Distributive Realities

Drugs on the market usually have an indication range for which they are officially registered and approved for the treatment of certain diseases by the responsible authorities (e.g., "Bundesamt für Arzneimittel und Medizinprodukte" in Germany or the Food and Drug Administration in the US). However, most of these drugs may also be beneficial for disorders for which the drug companies have no official approval, and doctors are also free to prescribe the drugs for other disorders, which is called *off-label medication.*

For example, drugs such as pramipexole or L-DOPA, which are approved for the treatment of Parkinson's disease, are frequently prescribed to people suffering from restless-legs syndrome. Likewise, ADHD may be treated *off-label* with modafinil, a drug that has been approved for the treatment of narcolepsy. Moreover, dementias are often treated with antidepressant drugs, e.g., SSRIs, in addition to cognitive enhancers. Hence, drugs that have been labeled "antipsychotics" may, for example, be of use for the treatment of mood disorders or addiction. The procedure of *off-label* medication is certainly supported by a paradigm shift in psychiatry and neurology, where recent years have seen a weakening of the nosological taxonomy of neuropsychiatric diseases and a strengthening of the dimensional classification of diseases, as predicted two decades ago (Crow 1990).

Considering the astonishing range of drugs that have relatively few side effects and effectively mitigate emotional and cognitive deficits in the course of illnesses, we are bound to ask ourselves whether or not we should take some of these drugs in order to improve these mental functions or emotional states in the *absence* of a diagnosed disease.[2,3]

8.5 Cosmetic Psychopharmacology? New Findings and Rising Expectations

The term "cosmetic psychopharmacology" was coined around 1990 by Peter Kramer, author of the book "Listening to Prozac". It implies using drugs to improve the quality of life rather than curing a disease, quite similar to plastic surgery or fertility medicine.

Mental health medicines cover at least three main domains of brain function: First, anxiolytic and antidepressant drugs in a broad sense can relieve patients from distinct fears (e.g., fear of heights, social or animal phobias) and from depressive mood, but also from more general states of anxiety, apprehension and concern. Second, psychoactive or psychostimulant drugs are used for recreational purposes, but also increase our vigilance, self-esteem and "power" and may be used to improve our professional performance. Third, cognitive enhancers mitigate cognitive deficits in several dementias and may be useful to increase our mental performance in general. Cognition is a heterogeneous construct encompassing perception and attention, learning and the representation of knowledge in the brain (short- and long-term memory, as well as its retrieval), reasoning, planning, and also social cognition. Hence, it is conceivable that drugs improving these mental core functions will also improve general well-being and happiness (Table 8.1).

---

[2] It is important here to note that it is not trivial to define "illness" and delimit this state from the healthy state. Definitions of diseases and concepts of treatment have changed over time and will continue to evolve. Likewise, it is already good practise to prescribe medications preventively in the *absence* of an illness, for example for vaccination, or consider the use of statins and antihypertensive drugs to reduce the risk of heart failure.

[3] It is actually hard to reliably tell how many people really use neuro-enhancement in everyday life. Applying a broad definition of neuro-enhancement, one might consider consuming coffee, tea or herbal extracts like *Ginkgo biloba*, smoking tobacco, and even meditation or hard exercise as neuro-enhancing activities (Hillman et al. 2008). Due to technical difficulties in running reliable surveys, there is very little consistent data on the incidence of use of prescription drugs as neuro-enhancers by healthy people. Most available information is based on anecdotal evidence or questionnaires. A journalist's survey among the readers of the science magazine *Nature* revealed an astounding finding, where about 20 % admitted having taken various drugs in order to improve their mental performance (Nature 452:674–675, 2008).

Table 8.1 Possible domains of neuro-enhancement

Brain- or mental function	Drug class	Example	Effect
Fear, anxiety, mood	Anxiolytics, antidepressive drugs	Benzodiazepines (e.g., valium), SSRIs (selective serotonin reuptake inhibitors)	Relief of fear, apprehension, concern mood stabilization
Vigilance, drive, alertness	Psychostimulants	Methylphenidate, Modafinil	Psychostimulant
Cognition (attention, learning, memory, executive functions)	Cognitive enhancers	Donepezil, Ampakines, phosphodiesterase inhibitors	Improvement of performance

About 80 years ago, Aldous Huxley wrote in his famous science fiction novel *Brave New World* about the drug "soma" that is taken by the members of the society as an aphrodisiac and in order to balance their mood (Huxley 1932). Only sixty years later, Huxley's utopia became quite real, when the drug company Eli-Lilly marketed the selective serotonin reuptake inhibitor Prozac (fluoxetine) for the treatment of depression. The "miraculous" effects of Prozac became known to a broad range of people when the psychiatrist Peter Kramer wrote a book reporting on the mood-balancing (and anorexic) effects of this drug in individual patients. Some of these patients claimed feeling even "better than well", paving the way for what Kramer called "cosmetic psychopharmacology", i.e. pills offering happiness for all (Kramer 1993). The mood-improving effects of Prozac are due to increasing the amount of the neurotransmitter serotonin in some parts of the brain.

Dopamine is another transmitter that has long been known to be involved in positive feelings and reward-related behavior. In the early days of brain stimulation experiments, Olds and Milner showed in rats that electrical stimulation of parts of the brain is so rewarding that rats would press the lever of a Skinner-box and self-stimulate their brains unceasingly until complete exhaustion (Olds and Milner 1954). Later, it turned out that the hot-spots in the brain eliciting massive reward—the so-called "pleasure centers of the brain" (Olds 1956)—are largely congruent with a part of the ascending dopaminergic system, which connects dopaminergic neurons in the ventral midbrain with the nucleus accumbens in the basal forebrain (Berridge 2007). Natural rewards and most addictive drugs act via a direct or indirect increase in the activity of the brain's reward system (Badiani et al. 2011; Barbano and Cador 2007; Bartels and Zeki 2000; Blood and Zatorre 2001, Breiter et al. 2001; Everitt and Wolf 2002; Kelley and Berridge 2002; Knutson et al. 2001; McClure and Montague 2004).

Interestingly, the increase in synaptic levels of the transmitter dopamine induced by psychostimulant drugs such as Ritalin not only activate the reward system of the brain but at the same time improve a variety of cognitive processes. Through projections from the midbrain to the frontal cortex, dopamine is also very important for the proper functioning of the cerebral cortex (Robbins 2000;

Winterer and Weinberger 2004). For example, it has been shown that Ritalin improves working memory and cognitive flexibility in healthy volunteers (Elliott et al. 1997).

However, not all people respond in a similar way to such drugs. Other work related to the above mentioned findings revealed an important caveat: Dopamine levels in different parts of the brain, e.g., in the frontal cortex, obviously have an *optimal* rather than a *maximal* level for perfect performance, which means that, depending on the difficulty of a mental task, neither too much nor too little dopamine is optimal (Chamberlain et al. 2006b; Dalley et al. 2004; Robbins 2000). Hence, a drug like Ritalin taken in a certain dose may improve cognitive functions in some, but not in other individuals, depending on their individual baseline levels of dopamine.

Individual differences in these baseline levels of dopamine are probably based on slight genetic differences—the so-called single nucleotide polymorphisms (SNPs). For example, it has been shown that SNPs in one of the dopamine metabolizing enzymes, catechol ortho-methyltransferase (COMT), lead to profound inter-individual differences in the cognitive enhancing effects of a psychostimulant. Those individuals with low dopamine levels benefit from psychostimulant treatment, whereas those people with high levels do not (Mattay et al. 2003). The immense progress made by molecular genetics in recent years and the subsequent linkage between the genome and the action of drugs has an important impact on the future of pharmacopsychiatry and on the potential of neuro-enhancement. It offers the possibility of detecting risk genes for certain disorders and for a personalized genotype-based drug treatment with minimal side-effects (O'Donovan et al. 2009).

Other drugs that are used to treat ADHD have also been found to increase cognitive functions in healthy subjects. For example, modafinil, a drug that enhances cholinergic and noradrenergic neurotransmission, has been licensed for the treatment of narcolepsy, but has also been shown to improve recognition memory, planning, and performance in several intelligence tests (Turner et al. 2003). Atomoxetine is a selective noradrenaline reuptake inhibitor that has a similar clinical profile to Ritalin and has also been found to improve cognition and executive functions in healthy humans (Chamberlain et al. 2006a).

Recent findings suggest a very important role for the neuropeptide oxytocin in social cognition and behavior (Kirsch et al. 2005). The levels of oxytocin and oxytocin receptors in the brain, in both animals and humans, have been found to be positively correlated with social affection, empathy, attachment and trust (Meyer-Lindenberg et al. 2011). Hence, the oxytocin system has been considered a future target for the pharmacotherapy of disorders characterized by deficits in social behavior, such as autism spectrum disorders. It is also conceivable it might be possible to increase sociability in healthy people, thereby improving one of the most important aspects of human life, namely societal integration.

Learning and memory are not only important cognitive functions for all organisms, but are essential for our autobiographical integration, professional success and happiness. Hence, the representation of knowledge in the brain is a core feature of human life. This is grimly reflected in the severe loss of contentedness seen in patients suffering from memory loss due to dementia or stroke. Cognitive enhancers in the domain of memory function so far have two major targets. First, the cholinergic system, where inhibitors of the enzyme acetylcholinesterase (e.g., donepezil or rivastigmin) improve cognitive functions, for example in Alzheimer's disease (Briand et al. 2007; Courtney et al. 2004).

Second, a more recent field of research emerged from the early findings in animal learning physiology, showing that the excitatory transmitter glutamate and its receptors are very important for learning and memory (Kandel 1991). It has been known since about the early 1970s that a phenomenon called "long-term potentiation" (LTP) might be a physiological correlate of learning (Malenka and Nicoll 1999). LTP was shown to depend on a particular glutamate receptor, the so-called NMDA receptor (this receptor was named after the compound N-Methyl-D-Aspartate, which strongly binds to this receptor). This receptor is crucial for the initial cellular and molecular events that are necessary for memory acquisition in neuronal systems.

However, there is another type of glutamate receptor, called AMPA-receptor (named after α-Amino-3-Hydroxy-5-Methyl-4-Isoxazolpropionic Acid, which selectively binds to this receptor) that is abundant on neurons in the central nervous system and co-localizes with NMDA-receptors in some regions of the brain. It has been found that activation of NMDA-receptors in the course of learning stimulates the synthesis, presence and activity of AMPA-receptors in memory systems, which is considered an essential aspect of long-term memory (Rensing et al. 2009; Collingridge et al. 2004).

Recently it has been shown that positive modulators of AMPA-receptors, the so-called Ampakines (e.g., CX-516), can boost their activity and, hence, improve memory encoding (Lynch 2002). Indeed, recent progress in the molecular neurobiology of learning and memory has yielded several further possibilities for memory improving drugs (Sacktor 2011).

Another avenue of memory research has been based mainly on the work of the Nobel prize winner Eric Kandel, who showed an important role of the protein CREB in long-term memory formation (Barco et al. 2003). The acronym CREB stands for cAMP-Response Element Binding protein, a nuclear transcription factor that changes the gene expression and phenotype of a neuron in the course of learning. Since CREB is activated by cyclic adenosine monophosphate (cAMP) it was suggested that phosphodiesterase inhibitors like MEM 1414 could enhance memory formation by increasing the cellular concentration of cAMP.

Given that a perfect memory is desirable for the retention of necessary knowledge, there are cases where forgetting and erasing unpleasant memories may be urgently wanted. For example, in trauma victims, where flashbacks are haunting people suffering from post-traumatic stress disorder, it may be necessary to get rid of the highly aversive memory traces in the brain (Shin et al. 2004). This line of reasoning has led researchers to develop a drug treatment that enhances the extinction of fear during exposure therapy in patients with fear of heights (Davis et al. 2006). During exposure therapy, patients are confronted in a controlled way with the specific fear-eliciting stimuli and/or contexts, so the fearful response can eventually be controlled by the patient. The drug D-Cycloserin significantly improved the outcome of exposure therapy in vertigo patients, and could probably also increase selective memory extinction in otherwise healthy people (Ressler et al. 2004).

8.6 Weighing the Pros and the Cons of Neuro-Enhancement

What are the costs—in terms of side effects—of these treatments and what are their benefits? One might argue that there are three basic prerequisites for taking medications as neuro-enhancers: First, they should not be unhealthy. Second, they should be taken autonomously and unforced. Third, there should be a reasonable degree of distributive justice in the society.

Psychopharmacological research and development strives hard to increase the beneficial power of drugs. As we have listed above, anxiolytic and antidepressant drugs improve well-being and mood. Psychostimulant drugs increase our vigilance, self-esteem and "power" and may improve our professional performance. Cognitive enhancers can be useful to increase our intellectual performance in general. Hence, neuro-enhancement drugs can improve mental core functions and well-being.

However, there are side effects. Brain research clearly shows that changing highly complex interacting networks of nerve cells by chronic drug use will inevitably lead to disturbances of the system. Recent findings indicate that long-term drug treatment might lead to changes not only in the "soft-wiring" of the brain, but can also induce permanent changes in gene activity that ultimately re-construct parts of the brain (Robinson and Nestler 2011). These side effects on the brain may be mild and tolerable, but they could also permanently impair functioning. In an already disturbed network—such as in the context of a mental disorder—the benefit induced by the drug is likely to be higher than the costs. It is also noteworthy that drugs may affect the brain differently during certain stages of life, for example during puberty. If we are to support neuro-enhancement in school-age children, we have to keep in mind that puberty is a maturational phase of the brain where it is very sensitive, for example, to the effects of cannabinoids (Caspi et al. 2005; Schneider and Koch 2003).

But what about the brain of a healthy adult human? Whilst the safety profiles of drugs that have been in clinical use for many years, such as Ritalin or Prozac, are known and the risk of side effects may be predictable, there are still little data on long-term use for newly developed drugs, e.g., the Ampakines. But still people might still consider the benefits higher than the costs.

Moreover, as mentioned above, health and disease are not that clearly defined, and the level of tolerance for unhealthy side effects of drugs is highly individual.

The fact that a drug is potentially dangerous for the individual who takes it justifies neither a moral nor a legal ban, since the individual is free to decide whether he or she takes the health risks (see, for example smoking or drinking alcohol).

Prerequisite number two is hard to control, as we are constantly under more or less subtle societal pressures (e.g., by fashion advertisements etc.). Certainly there are ethical concerns about neuro-enhancement or "brain doping" in order to maximize our mental capabilities if it becomes non-consensual by societal coercion after widespread use of such drugs. For example, if jobs would only be available to those who are willing to enhance their working power by drugs, or if drug intake would in any way violate individual autonomy.

Finally, concerning the distributive justice of such drugs, it has to be stated that many things are distributed unfairly in our society, mainly due to individual differences in socioeconomic status. However, one might argue that one should not further enhance the social discrepancies in our society by selectively marketing these drugs. But if the society is consensual on the issue of freely using certain drugs for neuro-enhancement, the question arises how this should be financed. We also have to consider that the term "equitably distributed" might mean "given to those who need it", rather than "completely free access to neuro-enhancers for all".

To conclude, there is quite a broad and still increasing range of drugs that can improve mental and emotional functions in ways to increase happiness. In the future, the possibilities of personalized drug-treatment based on individual genotype will probably increase the effectiveness of drugs and decrease their side effects. However, much further research is necessary to be clear about the risks and benefits of taking a particular drug. When health and personality risks are acceptable, the individual has the free choice to take a drug when it is quite likely that we will see a further increase in the *off-label* use of drugs in order to maximize happiness. Social thinking, moral philosophy and policy will have to integrate this into reasonable forms of regulation of neuro-enhancement (Greely et al. 2008; Singh and Kelleher 2010). If there is a principal societal agreement to use this technology, we would need special educational programs for general practitioners and other health care personnel in order to provide the necessary information to those willing to take neuro-enhancers.

Finally, we must not forget that the overall scheme of a happy and successful life cannot only be transfixed to mental functions like learning, memory, reward and drive. In that sense, neuro-enhancement can be regarded as a minor, supportive technology that may help us to achieve our plans and goals. We must be careful not to consider our lives as mere chemical processes that can be manipulated as desired (Farah 2012).

References

Badiani A, Bellin D, Epstein D, Calu D, Shaham Y (2011) Opiate versus psychostimulant addiction: the differences do matter. Nat Rev Neurosci 12:685–700

Barbano MF, Cador M (2007) Opioids for hedonic experience and dopamine to get ready for it. Psychopharmacology 191:497–506

Barco A, Pittenger C, Kandel ER (2003) CREB, memory enhancement and the treatment of memory disorders: promises, pitfalls and prospects. Exp Opinions Ther Targets 7:101–114

Bartels A, Zeki S (2000) The neural basis of romantic love. NeuroReport 11:3829–3834

Berridge KC (2007) The debate over dopamine's role in reward: the case for incentive salience. Psychopharmacology 191:391–431

Blood AJ, Zatorre RJ (2001) Intensely pleasurable responses to music correlate with activity in brain regions implicated in reward and emotion. Proc Nat Acad Sci USA 98:11818–11823

Breiter HC, Aharon I, Kahneman D, Dale A, Shizgal P (2001) Functional imaging of neural responses to expectancy and experience of monetary gains and losses. Neuron 30:619–639

Briand LA, Gritton H, Howe WM, Young DA, Sarter M (2007) Modulators in concert for cognition: modulator interactions in the prefrontal cortex. Prog Neurobiol 83:69–91

Caspi A, Moffitt TE, Cannon M, McClay J, Murray R, Harrington H et al (2005) Moderation of the effect of adolescent-onset cannabis use on adult psychosis by a functional polymorphism in the catechol-O-methyltransferase gene: longitudinal evidence of a gene x environment interaction. Biol Psychiatry 57:1117–1127

Chamberlain SR, Müller U, Blackwell AD, Clark L, Robbins TW, Sahakian BJ (2006a) Neurochemical modulation of response inhibition and probabilistic learning in humans. Science 311:861–863

Chamberlain SR, Muller U, Robbins TW, Sahakian BJ (2006b) Neuropharmacological modulation of cognition. Curr Opin Neurol 19:607–612

Collingridge GL, Isaac JTR, Wang YT (2004) Receptor trafficking and synaptic plasticity. Nat Rev Neurosci 5:952–962

Cooper JR, Bloom FE, Roth RH (2003) The biochemical basis of neuropharmacology. Oxford University Press, Oxford

Courtney C, Farrell D, Gray R (2004) Long-term donepezil treatment in 565 patients with Alzheimer's disease (AD2000): randomized double-blind trail. Lancet 363:2105–2115

Creese I, Burt DR, Snyder SH (1976) Dopamine receptor binding predicts clinical and pharmacological potencies of antischizophrenic drugs. Science 192:481–483

Crow TJ (1990) The continuum of psychosis and its genetic origins. Br J Psychiatry 156:788–797

Dalley JW, Cardinal RN, Robbins TW (2004) Prefrontal executive and cognitive functions in rodents: neural and neurochemical substrates. Neurosci Biobehav Rev 28:771–784

Davis KL, Charney D, Coyle JT, Nemeroff CB (2002) Neuropsychopharmacology: the fifth generation of progress. Lippincott Williams & Willkins, Philadelphia

Davis M, Myers KM, Chhatwal J, Ressler KJ (2006) Pharmacological treatments that facilitate extinction of fear: relevance to psychotherapy. NeuroRx 3:82–96

Elliott R, Sahakian BJ, Matthews K, Bannerjea A, Rimmer J, Robbins TW (1997) Effects of methylphenidate on spatial working memory and planning in healthy young adults. Psychopharmacology 131:196–206

Everitt BJ, Wolf ME (2002) Psychomotor stimulant addiction: a neural systems perspective. J Neurosci 22:3312–3320

Farah MJ (2012) Neuroethics: the ethical, legal, and societal impact of neuroscience. Annu Rev Psychol 10:571–591

Greely H, Sahakian B, Harris J, Kessler RC, Gazzaniga M, Campbell P, Farah MJ (2008) Towards responsible use of cognitive-enhancing drugs by the healthy. Nature 456:702–705

Hillman CH, Erickson KI, Kramer AF (2008) Be smart, exercise your heart: exercise effects on brain and cognition. Nat Rev Neurosci 9:58–65

Huxley A (1932) Brave new world. Chatto & Windus Ltd., Great Britain

Kandel ER (1991) Cellular mechanisms of learning and the biological basis of individuality. In: Kandel ER, Schwartz JH, Jessell TM (eds) Principles of neural science. Elsevier, Amsterdam, pp 1009–1032

Kelley AE, Berridge KC (2002) The neuroscience of natural rewards: relevance to addictive drugs. J Neurosci 22:3306–3311

Kirsch P, Esslinger C, Chen Q, Mier D, Lis S, Siddhanti S et al (2005) Oxytocin modulates neural circuitry for social cognition of fear in humans. J Neurosci 25:11489–11493

Knutson B, Adams CM, Fong GW, Hommer D (2001) Anticipation of increasing monetary reward selectively recruits nucleus accumbens. J Neurosci 21:1–5

Koch M (2007) On the effects of partial agonists of dopamine receptors for the treatment of schizophrenia. Pharmacopsychiatry 40:34–39

Kramer PD (1993) Listening to Prozac. Viking Penguin, New York

Loewenstein G, Rick S, Cohen JD (2008) Neuroeconomics. Annu Rev Psychol 59:647–672

Lynch G (2002) Memory enhancement: the search for mechanism-based drugs. Nat Neurosci 5:1035–1038

Malenka RC, Nicoll RA (1999) Long-term potentiation—a decade of progress? Science 285:1870–1874

Mattay VS, Goldberg TE, Fera F, Hariri AR, Tessitore A, Egan MF et al (2003) Catechol O-methyltransferase val^{158}-*met* genotype and individual variation in the brain response to amphetamine. Proc Nat Acad Sci USA 100:6186–6191

McCabe SE, Knight JR, Teter CJ, Wechsler H (2005) Non-medical use of prescription stimulants among US college students: prevalence and correlates from a national survey. Addiction 100:96–106

McClure SM, Montague PR (2004) The neural substrates of reward processing in humans: the modern role of fMRI. Neuroscientist 10:260–268

McKernan RM, Rosahl TW, Reynolds DS, Sur C, Wafford KA, Atack JR et al (2000) Sedative but not anxiolytic properties of benzodiazepines are mediated by the $GABA_A$ receptor α_1 subtype. Nat Neurosci 3:587–592

Meyer-Lindenberg A, Domes G, Kirsch P, Heinrichs M (2011) Oxytocin and vasopressin in the human brain: social neuropeptides for translational medicine. Nat Rev Neurosci 12:524–538

Olds J (1956) Pleasure centers in the brain. Sci Am 195:105–116

Olds J, Milner P (1954) Positive reinforcement produced by electrical stimulation of septal area and other regions of the brain. J Comp Physiol Psychol 47:419–427

O'Donovan MC, Craddock N, Owens MJ (2009) Genetics of psychosis; insights from views across the genome. Hum Genet 126:3–12

Platt ML, Huettel SA (2008) Risky business: the neuroeconomics of decision making under uncertainty. Nat Neurosci 11:398–403

Rensing L, Koch M, Becker A (2009) A comparative approach to the principal mechanisms of different memory systems. Naturwissenschaften 96:1373–1384

Ressler KJ, Rothbaum BO, Tannenbaum L, Anderson P, Graap K, Zimand E et al (2004) Cognitive enhancers as adjuncts to psychotherapy: use of D-cycloserine in phobic individuals to facilitate extinction of fear. Arch Gen Psychiatry 61:298–304

Robbins TW (2000) Chemical neuromodulation of frontal-executive functions in humans and other animals. Exp Brain Res 133:130–138

Robinson AJ, Nestler EJ (2011) Transcriptional and epigenetic mechanisms of addiction. Nat Rev Neurosci 12:623–637

Sacktor TC (2011) How does PHMζ maintain long-term memory? Nat Rev Neurosci 12:9–15

Schneider M, Koch M (2003) Chronic pubertal, but not adult chronic cannabinoid treatment impairs sensorimotor gating, recognition memory, and the performance in a progressive ratio task in adult rats. Neuropsychopharmacology 28:1760–1769

Shin LM, Orr SP, Carson MA, Rauch SL, Macklin ML, Lasko NB et al (2004) Regional cerebral blood flow in the amygdala and medial prefrontal cortex during traumatic imagery in male and female Vietnam veterans with PTSD. Arch Gen Psychiatry 61:168–176

Singh I (2008) Beyond polemics: science and ethics of ADHD. Nat Rev Neurosci 9:957–964

Singh I, Kelleher KJ (2010) Neuroenhancement in young people: proposal for research, policy, and clinical management. AJOB Neurosci 1:3–16

Smith ME, Farah MJ (2011) Are prescription stimulants "smart pills"? The epidemiology and cognitive neuroscience of prescription stimulant use by normal healthy individuals. Psychol Bull 137:717–741

Turner DC, Robbins TW, Clark L, Aron AR, Dowson J, Sahakian BJ (2003) Cognitive enhancing effects of modafinil in healthy volunteers. Psychopharmacology 165:260–269

von Bohlen und Halbach O, Dermietzel R (2002) Neurotransmitters and Neuromodulators. Wiley, Weinheim

Winterer G, Weinberger DR (2004) Genes, dopamine and cortical signal-to-noise ratio in schizophrenia. Trends in Neurosci 27:683–690

Chapter 9
Do Aspirations and Adaptation Impede the Maximization of Happiness?

Ulrich Schimmack and Hyunji Kim

9.1 Happiness as the Ultimate Goal

In this chapter, we use the term "happiness" to refer to a happy life rather than merely the experience of happiness and the absence of unhappy feelings. Our definition of happiness is close to Sumner's concept of authentic happiness (Diener et al. 2009; Sumner 1996). Happiness is a universal human concern. People who are happy want to remain happy and people who are unhappy want to be happy. Although much of human behavior can be explained by humans' attempts to maximize happiness, happiness is not a motive like other motives. That is, people have a pretty good idea what they need to do when they are thirsty, hungry, tired, too hot or too cold, threatened, or sick, but it is less evident what people should do to maximize their own happiness because happiness is defined as the match of an individual's actual life to a set of specific ideals. Thus, happiness requires the maximization of specific ideals (health, positive emotions, social relationships, etc.), which makes the pursuit of happiness more complex than the fulfillment of basic needs.

One important question about happiness is whether the active pursuit of happiness is possible and desirable. That is, should individuals reflect on their ideals, use these ideals to evaluate their lives, and actively try to close the gap between their actual and ideal life? Or is the active pursuit of happiness counterproductive because an active pursuit of happiness might ironically undermine happiness rather

U. Schimmack (✉) · H. Kim
Department of Psychology, University of Toronto Mississauga,
Mississauga, ON, Canada
e-mail: uli.schimmack@utoronto.ca

H. Kim
e-mail: hh.kim@utoronto.ca

U. Schimmack
Valerie Tiberius Department of Philosophy, University of Minnesota,
South Minneapolis, MN, USA

H. Brockmann and J. Delhey (eds.), *Human Happiness and the Pursuit of Maximization*, Happiness Studies Book Series, DOI: 10.1007/978-94-007-6609-9_9,

than increasing happiness? Philosophers, self-help books and happiness scientists give different and often contradictory advice. We review the contradictory claims about the benefits and costs of the pursuit of happiness in psychological literature.

9.2 Set-Point Theory: How the Pursuit of Happiness Undermines Happiness

One extreme view is that the very idea of pursuing happiness is foolish or a recipe for unhappiness. The main assumption underlying this view is that the causal factors that influence happiness are so complex and uncontrollable that individuals are unable to increase their chances of having a good life. As one prominent psychologist put it, you cannot pursue happiness; you can only stumble on it (Gilbert 2006). Accordingly, the best strategy to maximize happiness is to avoid pursuing happiness because happiness is an unattainable goal, and pursuing unattainable goals reduces happiness (Brunstein 1993). Scientific support for this position stems from Lykken and Tellegen's (1996) theory of happiness as a stochastic phenomenon. Accordingly, happiness is the result of two random processes, namely the random lottery of genetics and random events in the environment. Based on twin data, the authors concluded that there are genetic dispositions that make some people happier than others. Moreover, these genetic processes are so complex that even first-degree relatives who share 50 % of their genes are hardly any more similar to each other than two strangers. As a result, even happy parents cannot expect to have happy children, because children inherit their parents' genes, but not the complex interactions among them. As a result, only genetically identical twins have similar dispositions for happiness. Lykken and Tellegen (1996) acknowledge that life events can influence happiness. Everybody can recall good times and bad times in their lives. However, longitudinal twin studies suggest that these times never last for a long time, and that most of the stability in happiness is explained by genes.

Another finding that is consistent with the stochastic model of happiness is that average levels of happiness over the past three decades seem to have remained fairly stable despite improvements in living conditions, health care and longevity, and despite technological innovations such as the Internet (Easterlin 2000). This suggests that it is futile for governments to improve objective living conditions as a means to maximize well-being because improvements in these conditions have no influence on happiness. Consistent with this view, major literature reviews also suggest that objective life circumstances tend to have at best a small effect on happiness (Diener et al. 1999).

9.3 The Architecture of Happiness Theory: How to Pursue Happiness

The opposite view is that people are in charge of their own happiness. Accordingly, happiness can be generated by simply thinking happy thoughts. Happy thoughts not only make people feel good, they can also improve people's actual lives because positive thinking increases one's motivation, effective performance, and chances of success (Taylor and Brown 1988). Some scientists are a bit more reserved and suggest that happiness is not entirely under an individual's control. Seligman (2002) proposed that 50 % of the variance in happiness is genetically fixed and the other 50 % is due to life circumstances and other factors under people's control. Lyubomirsky et al. (2005) suggested that life circumstances explain only 10 % of the variance and that 40 % of the variance in happiness is due to factors that are controllable.

Somewhat surprisingly, this estimate is based on the same twin study that seemed to support the stochastic theory of happiness (Lykken and Tellegen 1996). To arrive at the 40 % estimate, Lyubomirsky et al. (2005) accept the assumption that genetic variance accounts for 50 % of the variance in happiness. They then use studies that regressed happiness measures on demographic variables to infer that 10 % of the variance is due to life circumstances that are not under people's voluntary control, which surprisingly also includes variables such as marital status. The 10 % estimate implies that genes and life circumstances account for only 60 % of the variance in happiness. The authors attribute the unexplained variance to factors that are under people's voluntary control. The main implication of this theory is that some people could be happier if they changed aspects of themselves or their lives that are under their control. Moreover, the 40 % estimate suggests that changes in these factors can produce notable increases in the happiness of individuals and populations.

To illustrate this potential, it is useful to think about happiness along a 0–10-point scale and to assume that genes, circumstances and voluntary factors independently contribute to happiness. To take the effect size of the three sets of factors into account, the contribution of each factor can be measured on a scale from 0 to 5 for genes, 0–1 for circumstances, and 0–4 for voluntary factors. An individual who is at the bottom for all three components would have a total score of 0. An individual who is at the top of all three components would have a total score of 10. If we assume normal distributions for all three factors, the average person would score a 2 on the voluntary factor scale. Thus, on average individuals could gain 2 points on the 0–10 happiness scale by fully maximizing their happiness potential. At the same time, this potential would be greater for individuals below average and smaller for individuals who are already doing things that maximize their happiness (i.e. if everybody maximized or failed to maximize happiness, there would be no variance that could be detected in a twin study). A 2-point increase in happiness corresponds to the difference in the average happiness of citizens in Romania (5.88) and the United States (7.73) (Seligman 2002). Thus, if happiness scientists could uncover the nature of the factors that are under people's control, it would be possible to make people in Romania as happy as US Americans and US Americans would be nearly perfectly happy (7.73 + 2 = 9.73 on a 0–10 scale).

Unfortunately, this promise rests on some questionable assumptions and interpretations of the scientific data. First, the estimates make the implicit assumption that happiness measures (often a single rating) are perfect measures of happiness. A more realistic assumption is that about 50 % of the variance is valid and the other 50 % of the variance reflects random and systematic measurement error (Schneider and Schimmack 2010). This could suggest that the true estimates for genes and circumstances are 80 % and 20 %, leaving hardly any residual variance that could be attributed to voluntary factors under people's control. Second, the estimate that life circumstances account for only 10 % of the variance is questionable because demographic factors capture only a fraction of the environmental factors that can influence well-being. For example, the small effect of marital status ignores the fact that the quality of the marital relationship also influences happiness (Lucas et al. 2003). That is, the effect of marital status only takes into account that, on average, married individuals tend to be somewhat happier than unmarried individuals. It does not take into account that being married to an abusive spouse or marital conflict can dramatically lower well-being (Lucas 2005).

It is also questionable to divide variance components into those that are controllable and those that are not controllable. For example, are life circumstances such as unemployment and marital status controllable or uncontrollable? It seems that they are only partially under voluntary control. For example, with an unemployment rate of 9 %, it is impossible for everybody to find a job, but individuals who are more motivated and determined have a better chance of getting a job.

Even genetic variance does not imply that these factors are fixed. For example, skin color and other ethnic features are genetically determined. In a society that discriminates against people with specific genetically determined features, this effect of race and discrimination would contribute to the genetic variance. However, societal factors such as discrimination can be changed and would reduce the genetic variance and increase the well-being of individuals who suffer from discrimination. Indeed, the happiness of African Americans in the United States has increased over the past decades since the civil rights movement (Diener et al. 2009). Even at the individual level, individuals who face discrimination may have some control over their happiness if they are able to move to a place where they face less discrimination or if they face discrimination in the workplace and are able to switch jobs to a company with a better work environment.

Finally, Lyubomirsky et al. (2005) fail to take the longitudinal evidence into account. Accordingly, most (80 % of the stable variance) can be attributed to genetic factors. Some stable life circumstances may account for another 10 % of the stable variance, leaving only 10 % of the stable variance for voluntarily controllable factors. One might argue that some of the voluntary controllable factors change over time, but this would imply that some people stop doing activities that make them happy and that are under their voluntary control. This seems unlikely because happiness is reinforcing and people are likely to continue doing things that make them happy. In sum, positive psychologists' optimistic view about the chances of dramatic increases in happiness rests on questionable assumptions and interpretations of scientific findings.

9.4 What is the Real Maximum of Happiness?

A more balanced account of the scientific evidence suggests that the pursuit of happiness is neither futile nor a cakewalk. Ironically, one problem for many people to increase their happiness (further) is that most people are quite successful in their pursuit of happiness (Diener and Diener 1996). As a result, they are close to the maximum level of happiness that they can achieve given their actual personality, preferences and life circumstances.

The average level of happiness in wealthy, democratic nations with good health care in Northern Europe, North America and Australia is close to 8 on a 0–10 scale (Deaton 2008). A value of 8 out of 10 seems to suggest that there is still room for improvement. However, it is unrealistic to assume that all people can be perfectly happy all the time. First, twin studies suggest that stable dispositions produce stable variation in happiness and variation implies that some individuals are below the maximum level. If genes contribute 50 % of the variance and are normally distributed (due to random recombination of genes from one generation to the next), the average genetic potential would translate into a score of 2.5 on a 0–5 scale. This would imply a maximum of 7.5, even if all other factors were ideal. The fact that the actual value is higher already shows that this model is too simplistic, but it does illustrate the implications of stable genetic dispositions for individuals' and populations' maximum values.

It is also noteworthy that there are dramatic differences in the average level of happiness across nations (Deaton 2008). This has two implications. First, twin studies overestimate the effects of genetic factors because they estimate heritability within nations with a restricted range of living conditions. If we take the full range of living conditions into account, genes may only account for 30 % of the variance, rather than 50 %, which in my simplistic model would translate into an average of 1.5 on the 0–3 scale for genetic factors and a theoretical maximum of 8.5 if all other factors were ideal. The actual averages in the happiest nations are close to this maximum.

National variation in happiness also suggests that good living conditions in the happiest nations contribute to the happiness of these nations. Thus, an average score of 8 out of 10 is close to the maximum that can be reached with an average genetic influence of 1.5 and a perfect score for environmental factors (7 on a 0–7 scale; $7 + 1.5 = 8.5$).

This interpretation of the data would suggest that Lykken and Tellegen (1996) were right to tell North Americans that trying to be happier is futile, but that this conclusion does not generalize to individuals in other nations such as Romania, India or many other nations with less ideal circumstances. Moreover, the reason would not be that environmental factors are unimportant. The reason would be that the environmental factors in the US are close to the maximum of its happiness potential. It is simply unrealistic to assume that the average could increase further by a full point or two.

A more realistic view suggests that increases in national averages of happiness will be relatively small, but this does not mean that they are unimportant. Unfortunately, psychologists are not trained to think about effect sizes, or they focus on standardized effect sizes. According to (Cohen 1992) influential guidelines, a small effect is a difference of 0.1 standard deviation. The standard deviation on a 0–10 happiness rating is about 1.5. Thus, a mean difference of 0.15 would be a small effect. It is also a convention in psychology to report results to the second decimal place. Thus, a mean difference of 0.004 would be reported as if there was no change. Moreover, sample sizes tend to be very small to show that changes are not just random. A power analysis shows that 78,491 participants would be needed to demonstrate that a change by 0.01 units on a 0–10 scale (and SD = 1.5) is statistically significant. This is important because even a major national crisis has a muted effect on national averages of happiness that can be difficult to detect in existing data. This problem is even more severe when happiness is being measured on a 3-point scale as in the General Social Survey in the United States, which makes it extremely difficult to detect changes in national averages.

To illustrate this problem, we use the recent economic crisis in the United States as an example. As a result of this crisis, unemployment increased dramatically from around 5 to 10 %. This increase in unemployment is expected to produce a decrease in happiness because studies of individuals show that unemployment produces a notable decrease in happiness (Lucas et al. 2004). However, it is unlikely that current measures of happiness can show this effect at the level of national averages. Assuming that individual unemployment would result in a 1-point drop on the 0–10 scale (a moderate to strong effect of about 2/3 of a standard deviation), an increase in the unemployment rate by 5 % points would produce a decrease in the national average by 0.05 points. This would translate into a standardized effect size of $d = 0.03$. Moreover, the actual decrease would be smaller because national averages include individuals who do not participate in the workforce. It is therefore not surprising that daily surveys of happiness in the United States show no clear effect of the unemployment rate on average levels of happiness (http://www.well-beingindex.com/default.asp). It is therefore more informative to examine the factors that influence happiness at the individual level.

9.5 Adaptation Theory: Does Hedonic Adaptation Undermine Maximization of Happiness

The most popular explanation for suboptimal levels of happiness is adaptation level theory, which is also often called "hedonic treadmill theory" (Brickman et al. 1978; Diener et al. 2006). The main premise of adaptation level theory is that happiness reflects appraisals of events in comparison to a standard of comparison that is called an "adaptation level and not ." Adaptation levels shift in accordance with changes in actual life circumstances. When living conditions improve, adaptation levels increase. When living conditions get worse, adaptation levels decrease. As a result, happiness does not reflect how good people's lives are, it merely reflects whether life is getting better or worse.

Although adaptation level theory creates problems for increasing happiness, it also has positive implications for happiness when life is getting worse. Thanks to a lower adaptation level, people in objectively bad circumstances are as happy as those in more fortunate circumstances. Prolonged periods of unhappiness should only occur when life conditions continue to deteriorate.

Despite its popularity, adaptation theory faces numerous theoretical and empirical challenges. A theoretical problem for adaptation theory is that hedonic adaptation is often used to describe temporal patterns in happiness data without a theoretical account of the underlying causal mechanisms. Even if it is not entirely circular, it is confusing to use the term *adaptation* to describe a return of happiness levels to the level before an event and to attribute this phenomenon to a process called *adaptation* (Hahn 2011). The actual processes underlying the phenomenon often remain unspecified.

Hedonic adaptation is often compared to sensory adaptation. However, this analogy may be misleading. One example of sensory adaptation is how perception adapts to different light conditions when people leave a dark house and step into bright sunlight or vice versa. After a while, the sensory system adapts to the new conditions to allow optimal functioning in these conditions. As a result, vision is always at the optimal (i.e. maximum) level that particular light conditions afford. If hedonic adaptation functioned in the same manner, it would ensure that humans were always at the optimal level of happiness. However, this prediction is inconsistent with the observation that happiness is highest during times of positive change and that hedonic adaptation is invoked to explain suboptimal levels of happiness. The fact that we can immediately feel happy about an event suggests that the evaluation is based on some standard that has nothing to do with the current situation. For example, we may feel pleasantly surprised rather than annoyed when our work gets interrupted by a call from a friend. Although this phone call disturbs the equilibrium, it can elicit happiness because it is appraised with regard to some other goal (e.g., valuing or maintaining friendships). Given the complexity of the causal processes underlying happiness and other emotions, using sensory adaptation as an analogy is questionable.

The complexity of human emotions also suggests that adaptation theory may be limited to some emotions. Approach-motivated positive emotions such as excitement may signal goal progress. Moreover, people are likely to adjust their goals in response to their achievements. For example, a new world record sets a new and higher standard for future athletes. Thus, brief periods of excitement may not contribute to lasting happiness (Diener et al. 1991). However, other hedonic experiences such as contentment or sadness may be less susceptible to adaptation processes.

Finally, adaptation theory implies that life circumstances are fairly stable so that people have time to adapt to them. However, lives are more complex than bright or dark rooms. Life is constantly changing and human emotions respond to these changes. This is important because adaptation requires constant stimuli and does not apply to repeated events. For example, for many people eating chocolate is pleasurable, yet the pleasure diminishes with increased consumption and can even turn into displeasure at some point. However, most people who enjoy eating chocolate eat just enough to enjoy it. When it is not enjoyable, they stop, but then enjoy it again after some time. Billion-dollar sales of chocolate, pizzas, condoms and other consumer goods suggest that hedonic adaptation does not apply to repeated events. Commuting remains less pleasant than having sex even when people have repeated these experiences many times (Kahneman et al. 2004).

Another empirical challenge comes from the large national differences in happiness (Deaton 2008). A radical adaptation theory suggests that people can get used to all life circumstances. Thus, there should be no national differences in happiness (Easterlin 1974). However, even sensory adaptation has limits. Similarly, one would expect that hedonic adaptation has limits and that it is easier to get used to some living conditions than others. The large national differences in happiness clearly demonstrate some limits to adaptation processes (Deaton 2008; Diener et al. 1995; Inglehart et al. 2008).

Some alleged support for adaptation theory rests on misinterpretation of empirical findings. For example, weak correlations between income and happiness have been interpreted as evidence for hedonic adaptation (Lucas and Schimmack 2010). Once more this interpretation is based on a misunderstanding of effect sizes. According to Cohen, a correlation of $r = 0.15$ is a weak effect, but he was considering normally distributed variables. On a normally distributed variable, it is difficult to increase by more than 6 standard deviations. However, income is not normally distributed and it is possible to earn 10 standard deviations more than the average person. The same linear correlation would imply that this produces an increase in happiness of 1.5 standard deviations. In reality, the wealthy are about one standard deviation above the happiness of the average US American (Diener et al. 1985). Moreover, in poorer nations, where income is a better indicator of life circumstances, the correlation is notably stronger (Diener and Oishi 2000).

There is disagreement in the literature about the causal mechanisms underlying this relationship. Absolute theories propose that money is a valuable resource for closing the gap between one's actual and ideal life. Relative theories propose that income matters because people make social comparisons with others (Easterlin 1974). Although adaptation theory is also a relative theory, it assumes that happiness is relative to past personal experiences. In contrast, social comparison theory does not predict adaptation to a set point. As income differences are quite stable (Luhmann et al. 2011), rich people would remain happier than poor people, if social comparison theory were correct.

The use of correlations and related effect size measures such as amount of explained variance is even more problematic for studies that examine the effects of life events on happiness. The reason is that correlations and amount of explained variance focus on the contribution of a variable to the variance in happiness. This is problematic because the contribution of the variance in life circumstances to the variance in happiness is a function of two independent factors, namely the amount of variance in these objective life circumstances and the effect size of these circumstances on the happiness of individuals. The sensitivity to the variance in circumstances makes correlations poor indicators of effect sizes for rare events. For example, Diener et al. (2006) cite a correlation of $r = 0.08$ between objective health and happiness as evidence for adaptation theory. This is problematic because the correlation may be small for the simple reason that most people are fairly healthy. As a result, even large effects of poor health on happiness cannot produce strong correlations. In fact, Lucas (2007) demonstrated that severe disability has a strong and lasting negative effect on happiness. Other studies also show that poor health is a negative predictor of happiness and that the high well-being of older individuals in wealthy nations can be partially attributed to the high portion of GDP that is spent on the health care of older citizens, whereas poor health contributes to a negative correlation between age and happiness in poorer nations (Deaton 2008).

Another methodological problem is that adaptation theory predicts the null hypothesis and that researchers often make the mistake of interpreting a non-significant correlation as evidence for the null hypothesis. This inference is usually not warranted because small samples do not provide sufficient statistical power to distinguish between the null effect predicted by adaptation theory and a small effect that contradicts adaptation theory. For example, a study by Suh et al. (1996) has been cited as evidence that life events do not have effects that can last more than three months, suggesting that adaptation is a fast process (Diener et al. 2006). The actual results showed correlations of $r = 0.25/-0.28$ ($p < 0.05$) for events in the past three months ($p < 0.05$), and $r = 0.16/-0.12$ ($p > 0.05$) for events six months ago (Suh et al. 1996). However, given the modest sample size of this study (N = 115), these correlations are also not significantly different from each other, as indicated by the overlapping 95 % confidence intervals (0.25, 0.11|0.39 vs. 0.16, 0.01|0.31 and −0.28, −0.42|−0.14 vs. −0.12, −0.27|0.03). Thus, the results do not show that events in the past three months are significantly different from events in the past six months.

Diener et al. (2006) furthermore note that most of the events were rather mundane. The most common positive event in this student sample was getting an A for a college course. It is therefore not surprising that studies with larger samples and more important life events such as unemployment, divorce and widowhood show that adaptation is a slow process that can take several years and does not always result in complete adaptation (Lucas 2005).

Another problem in studies that claim support for adaptation theory is that events are often classified as good or bad based on some social norms, but that happiness responds to the subjective appraisal of these events, which can diverge from these social norms. For example, Suh et al. (1996) list marriage of a sibling as a positive event. However, this event could be appraised negatively and lower happiness because the event could imply that a close sibling relationship will become less intimate.

Adaptation theory also does not take into account that the appraisal of events can change over time. A romantic break-up may be appraised as a loss at first, but it may be appraised positively when the individual finds a new partner (Lucas 2005). Although these changes in appraisals make it difficult to predict how life circumstances influence happiness, changing appraisals would still imply that these circumstances can have long-lasting effects.

Yet another problem is that happiness researchers often have to rely on correlational data to make inferences about causality. A common statistical approach with correlational data is to run multiple regression models. This can lead to an underestimation of the importance of life events if different life events are causally related to each other. For example, distant events could have an indirect effect on happiness that is mediated by more recent events. A simple regression model would fail to show this effect. For example, one study found that childhood sexual abuse was a predictor of lower happiness in adulthood and that this relationship was no longer statistically significant after including more recent sexual experiences in the model (Browning and Laumann 1997). The lack of a significant effect in the multiple regression model does not demonstrate that childhood sexual abuse has no long-term effect on happiness. An alternative interpretation of the results is that childhood sexual abuse has negative effects on sexual pleasure in adulthood, which in turn reduces happiness in adulthood. Alternative interpretations of the data are possible, but the main point is that it is difficult to make inferences about the effects of life events on happiness without taking into account that life events have multiple short-term and long-term consequences on other aspects of people's lives.

Another common fallacy in the interpretation of research findings is to interpret patterns in means without taking variation in these means into account. For example, a longitudinal study of marriage and life satisfaction is often cited as evidence for adaptation theory. "The treadmill model of happiness posited by Brickman and Campbell (1971) represents a milestone in psychologists' understanding of happiness, and our longitudinal findings on marriage support the treadmill idea" (Diener et al. 2006, p. 312). This conclusion is based on the observation that the average level of life satisfaction two years after marriage was the same as the average level of life satisfaction before marriage. Yet, the authors of the original article also examined the effect of marriage at the level of individuals. Here the results do not show adaptation and do not support the treadmill idea. Specifically, the results suggested that some individuals had lasting gains in life satisfaction after marriage, whereas others had lasting declines in life satisfaction. The latter finding is also consistent with the observation that low levels of life satisfaction precede divorce (Lucas 2005). This finding implies that individuals do not simply adapt to their spouses. In conclusion, hedonic adaptation is often seen as a powerful force that impedes the pursuit of happiness, but the evidence for adaptation as a quick, automatic process is much weaker than popular reviews of the literature suggest.

9.6 Revised Adaptation Theory (RAT)

To address some of the limitations of adaptation theory, Diener et al. (2006) revised it. The first revision concerns the average hedonic tone during a state of complete adaptation. The original theory implied that hedonic adaptation results in a state of indifference or neutrality (neither good nor bad). Diener et al. (2006) point out that survey data from around the world consistently find that average levels of happiness are above a neutral point. Even in a relatively neutral laboratory setting, people do not report feeling indifference (neither pleasure nor displeasure), but mild levels of pleasure (Schimmack 2001). To account for this finding, Diener et al. (2006) propose a positive set point. Although this is one possible explanation of the evidence, it is difficult to create neutral situations to measure the hedonic tone of set points. For example, people's feelings are also influenced by internal bodily processes such as bodily pain, hunger, temperature and sleep. A mild positive affect may be positive feedback that these bodily systems are at or near their optimal level. It is likely that prolonged states of hunger, tiredness or pain are associated with displeasure. When affect is measured in naturalistic settings, it is even more likely that happiness levels are not just the result of a passive adaptation process, but that people actively contribute to their happiness. For example, people seek out the company of friends rather than enemies, they burn fossil fuels to maintain optimal temperatures in their homes, and they listen to music or talk on the phone to make commuting more pleasurable. In sum, a positive offset provides one possible explanation for the prevalence of happiness, but it is also possible that humans voluntarily shift their hedonic level towards the positive end.

Diener et al. (2006) note a second problem with the original adaptation theory. If adaptation were a universal process and events only produced temporary deviations from a set point, all individuals should have more or less the same level of happiness. Yet, there is ample evidence that stable dispositions produce variation in happiness across individuals (Ehrhardt et al. 2000; Fujita and Diener 2005; Schimmack and Lucas 2010), and twin studies suggest that these dispositions are inherited (Lykken and Tellegen 1996). Although the incorporation of dispositions into adaptation theory is a step in the right direction, the revised model still maintains the core assumption of adaptation theory that happiness reverts to a baseline level. It just allows for variation in this baseline level across individuals. However, dispositions can also interact with life events. For example, a depressive disposition may only trigger severe depression in response to major stressful life events (Caspi et al. 2003). Similarly, Oishi and Schimmack (2010) found that introversion was more detrimental for happiness for children who had moved around a lot. Personality can even interact with stable cultural factors. For example, it appears as if extraversion is a weaker predictor of happiness in Germany than in the United States (Schimmack et al. 2008).

Diener et al. (2006) further revised set-point theory by postulating multiple set points. They propose that multiple set points are needed to account for the fact that different happiness indicators can move in divergent directions. For example, the authors suggest different set points for positive affect and housing satisfaction because these two indicators moved in opposite directions. However, a test of set-point theory of housing satisfaction shows very little evidence for a housing-satisfaction set point (Nakazato et al. 2011). The authors in this study examined housing satisfaction before and after moving using a two-intercept growth model. One important finding was that the average level of housing satisfaction increased after moving and that this increase was maintained over a period of five years. Thus, the study does not support the hypothesis of adaptation theory that housing satisfaction quickly returns to a housing-satisfaction set point. The second important finding was that there was relatively low stability in the rank order of housing satisfaction before and after moving. This finding undermines the notion that differences in housing satisfaction are strongly influenced by variation in a stable housing-satisfaction set point. Finally, some of the stability in housing satisfaction before and after moving was explained by top-down effects from general happiness to housing satisfaction. This suggests that it is not necessary to postulate multiple set points for different happiness indicators. A more parsimonious explanation is that multiple happiness indicators are influenced by a common stable disposition. In fact, a general disposition seems to account for the fact that individuals who are satisfied with one life domain are also more satisfied with other domains (Schneider and Schimmack 2010).

9.7 Conclusion: The Pursuit of Happiness as Managing a Complex System

In conclusion, the notion of adaptation theory is misleading as it fails to elucidate the complex processes that shape lives and ideals. It has also created a false sense of uncontrollability. Major life events such as divorce can have prolonged effects on happiness, and adaptation is an active process that often requires finding a new partner rather than simply adapting to a life without one (Lucas 2005). Given the complexity and uncertainty about the hedonic consequences of major life decisions, the pursuit of happiness can be compared to the operation and management of a complex system. In a complex system, every action can have intended positive effects and unintended negative consequences. Moreover, complex systems constantly change and evolve and strategies that were successful in the past may no longer work in the future. Complex systems are also subject to unforeseen events that make it difficult to maximize rewards. If life were as simple as a Skinner box, it would not be so hard to maximize rewards and minimize punishments. However, life is more like a Skinner box with 1,000 levers with constantly changing reinforcement schedules. For example, buying a better house does improve housing satisfaction, but it may also have other negative effects making it difficult to predict whether there will be a net gain in happiness (Nakazato et al. 2011).

Similarly, marriage brings some rewards, but also new problems, and it can be difficult to foresee the net effect on happiness (Lucas et al. 2003). The general law of diminishing returns implies that it becomes increasingly more difficult to maximize happiness when happiness is already high. Another problem is that active pursuit of goals is effortful. Making the right choices can take a lot of time and can be frustrating (e.g., finding the best air fare). Having more disposable income often means working harder. Getting a great job may imply moving away from friends and family. Moreover, affluence creates new problems of choice: Would I be happier if I bought a vacation home in Florida or not?

The main goal of happiness science is to uncover some general principles that people can consider in their pursuit of happiness. One problem is that happiness science has few generally agreed facts and no overarching theory that explains these facts. However, even if this were to change, it would still be difficult to apply these theories to individuals' lives. For example, should an unemployed introvert move away from friends and family to get a job? The net effect on happiness will depend on so many specific factors that it will be difficult to forecast the effects on a single individual's happiness. At the same time, happiness science has produced some reliable and important findings. For example, there is clear evidence that some nations provide better opportunities for the maximization of happiness than others. Unfortunately, these nations also use a large amount of resources to do so. The scarcity of resources such as energy, food and water is just another reason why it is difficult to maximize human happiness for a world population of 9 billion people. To contribute to the management of human happiness, happiness science needs to create more sensitive measures, establish a set of uncontroversial empirical facts, and develop theories that can explain these facts.

References

Brickman P, Campbell DT (1971) Hedonic relativism and the good society. In: Appley MH (ed) Adaptation-level theory: a symposium. Academic Press, New York, pp 287–302

Brickman P, Coates D, Janoff-Bulman R (1978) Lottery winners and accident victims: is happiness relative? J Pers Soc Psychol 36(8):917–927

Browning CR, Laumann EO (1997) Sexual contact between children and adults: a life course perspective. (Article; Proceedings Paper). Am Sociol Rev 62(4):540–560

Brunstein JC (1993) Personal goals and subjective well-being: a longitudinal study. J Pers Soc Psychol 65(5):1061–1070

Caspi A, Sugden K, Moffitt TE, Taylor A, Craig IW, Harrington H et al (2003) Influence of life stress on depression: moderation by a polymorphism in the 5-HTT gene. Science 301(5631):386–389

Cohen J (1992) A power primer. Psychol Bull 112:155–159

Deaton A (2008) Income, health, and well-being around the world: evidence from the Gallup World Poll. J Econ Perspect 22(2):53–72

Diener E, Diener C (1996) Most people are happy. Psychol Sci 7:181–185

Diener E, Oishi S (2000) Money and happiness: income and subjective well-being across nations. In: Diener E, Suh EM (eds) Culture and subjective well-being. The MIT Press, Cambridge, pp 185–218

Diener E, Diener M, Diener C (1995) Factors predicting the subjective well-being of nations. J Pers Soc Psychol 69:851–864
Diener E, Horwitz J, Emmons RA (1985) Happiness of the very wealthy. Soc Indic Res 16:263–274
Diener E, Lucas RE, Scollon CN (2006) Beyond the hedonic treadmill: revising the adaptation theory of well-being. Am Psychol 61(4):305–314
Diener E, Sandvik E, Pavot W (1991) Happiness is the frequency, not the intensity, of positive versus negative affect. In: Strack F, Argyle M, Schwarz N (eds) Subjective well-being: an interdisciplinary perspective. International series in experimental social psychology, vol 21. Pergamon Press Inc, Elmsford, pp 119–139
Diener E, Lucas RE, Schimmack U, Helliwell J (2009) Well-being for public policy. Oxford University Press, New York
Diener E, Suh EM, Lucas RE, Smith HL (1999) Subjective well-being: three decades of progress. Psychol Bull 125(2):276–302
Easterlin R (1974) Does economic growth improve the human lot? Some empirical evidence. In: David PA, Reder MW (eds) Nations and households in economic growth: essays in honour of Moses Abramovitz. Academic Press, New York, pp 89–125
Easterlin RA (2000) The worldwide standard of living since 1800. J Econ Perspect 14:7–26
Ehrhardt JJ, Saris WE, Veenhoven R (2000) Stability of life-satisfaction over time: analysis of change in ranks in a national population. J Happiness Stud 1(2):177–205
Fujita F, Diener E (2005) Life satisfaction set point: stability and change. J Pers Soc Psychol 88:158–164
Gilbert DT (2006) Stumbling on happiness. Knopf, New York
Hahn U (2011) The problem of circularity in evidence, argument, and explanation. Perspect Psychol Sci 6(2):172–182
Inglehart R, Foa R, Peterson C, Welzel C (2008) Development, freedom, and rising happiness: a global perspective (1981–2007). Perspect Psychol Sci 3(4):264–285
Kahneman D, Krueger AB, Schkade DA, Schwarz N, Stone AA (2004) A survey method for characterizing daily life experience: the day reconstruction method. Science 306(5702):1776–1780
Lucas RE (2005) Time does not heal all wounds: a longitudinal study of reaction and adaptation to divorce. Psychol Sci 16(12):945–950
Lucas RE (2007) Long-term disability is associated with lasting changes in subjective well-being: evidence from two nationally representative longitudinal studies. J Pers Soc Psychol 92(4):717–730
Lucas RE, Schimmack U (2009) Income and well-being: how big is the gap between the rich and the poor? J Res Pers 43:75–78
Lucas RE, Clark AE, Georgellis Y, Diener E (2003) Reexamining adaptation and the set point model of happiness: reactions to changes in marital status. J Pers Soc Psychol 84(3):527–539
Lucas RE, Clark AE, Georgellis Y, Diener E (2004) Unemployment alters the set point for life satisfaction. Psychol Sci 15(1):8–13
Luhmann M, Schimmack U, Eid M (2011) Stability and variability in the relationship between subjective well-being and income. J Res Pers 45:186–197
Lykken D, Tellegen A (1996) Happiness is a stochastic phenomenon. Psychol Sci 7(3):186–189
Lyubomirsky S, Sheldon KM, Schkade D (2005) Pursuing happiness: the architecture of sustainable change. Rev Gen Psychol 9(2):111–131
Nakazato N, Schimmack U, Oishi S (2011) Effect of changes in living conditions on well-being: a prospective top-down bottom-up model. Soc Indic Res 100:115–135
Oishi S, Schimmack U (2010) Residential mobility, well-being, and mortality. J Pers Soc Psychol 98(6):980–994
Schimmack U (2001) Pleasure, displeasure, and mixed feelings: are semantic opposites mutually exclusive? Cogn Emot 15:81–97
Schimmack U, Lucas RE (2010) Environmental influences on well-being: a dyadic latent panel analysis of spousal similarity. Soc Indic Res 98:1–21

Schimmack U, Schupp J, Wagner GG (2008) The influence of environment and personality on the affective and cognitive component of subjective well-being. Soc Indic Res 89:41–60

Schneider L, Schimmack U (2010) Examining sources of self-informant agreement in life-satisfaction judgments. J Res Pers 44(2):207–212

Seligman MEP (2002) Authentic happiness: using the new positive psychology to realize your potential for lasting fulfillment. Free Press, New York

Suh E, Diener E, Fujita F (1996) Events and subjective well-being: only recent events matter. J Pers Soc Psychol 70(5):1091–1102

Sumner LW (1996) Welfare, happiness, and ethics. Clarendon Press, Oxford

Taylor SE, Brown JD (1988) Illusion and well-being: a social psychological perspective on mental health. Psychol Bull 103(2):193–210

Chapter 10
My Car is Bigger than Yours: Consumption, Status Competition, and Happiness in Times of Affluence

Hilke Brockmann and Song Yan

10.1 The Economics of XXL-Consumption

Before the meltdown of the financial industry in 2008, economic explanations and predictions were widely trusted. Economists seemed the best at comprehending what moves producers and consumers. Size and choice appeared to matter. From the 1990s onwards, rich market economies grew strongly, and private consumption followed suit. Correspondingly, low income countries, with their restricted markets, experienced slow growth. Figures 10.1 and 10.2 illustrate the steady rises in the rich world.

In neoclassical thinking, consumption reflects the needs and wants of consumers. Rising consumption is driven by the innate human desire to maximize utility. Based on this maximization assumption, Samuelson's (1938) theory of revealed preferences derives peoples' likes and dislikes from their shopping behavior. However, from this perspective the tastes and values of consumers remain independent of market demand and supply. A theory of preferences is no longer an economist's business. Consumers appear as sovereign actors. And "consumption, no matter how idiosyncratic, was viewed as the creator of demand and the motive for producers to create goods". Thus, consumption seems "an individual end in itself" (Frenzen et al. 1994, p. 405).

In 1977 Stigler and Becker elaborated on a new theory of consumer choice. By turning the consumer from "a passive maximizer of the utility from market purchases into an active maximizer also engaged in extensive production and investment activities" (Stigler and Becker 1977, p. 77), they completely abandoned the theory of tastes and values. From their point of view, preferences cannot be read from people's purchases because consumers do not have a taste for certain

H. Brockmann (✉) · S. Yan
School of Humanities and Social Sciences, Jacobs University Bremen, Bremen, Germany
e-mail: h.brockmann@jacobs-university.de

H. Brockmann and J. Delhey (eds.), *Human Happiness and the Pursuit of Maximization*, Happiness Studies Book Series, DOI: 10.1007/978-94-007-6609-9_10,

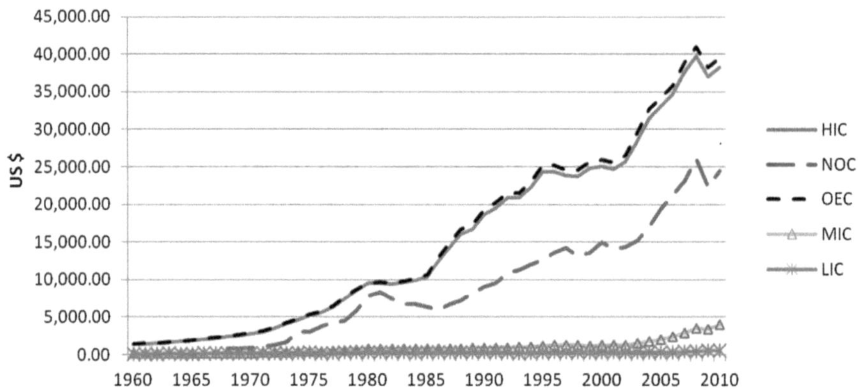

Fig. 10.1 Global economic growth in high, medium, and low income countries since 1960 (in GDP per capital and current US$). *Note*: According to the World Bank we divide economies according to 2011 GNI per capita into *HIC* (High Income Countries >=$12,475), *MIC* (Middle Income Countries $1,026–12,475), *LIC* (Low Income Countries <=$1,026), *NOC* (No OECD Countries), *OEC* (OECD Countries) *Source*: OECD National Accounts 2012

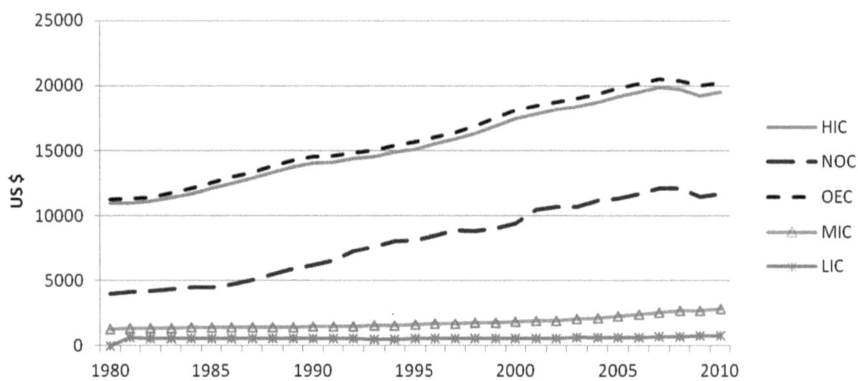

Fig. 10.2 Household consumption in high, medium and low income countries since 1980 (per capita and in current US$). *Source* OECD 2012

products. Instead, "tastes" refers to higher order goods like comfort or appreciation. By means of any purchasable consumer item or service, households manufacture higher order goods by themselves. Therefore prices and incomes are sufficient to explain consumer choices. Consequently, Stigler and Becker entitle their paper "De gustus non est disputandum". But the theoretical elegance and parsimony of the economic model comes at a price. The functional explanation of any consumption remains anemic and indeterminate. Its predictive power is weak. Everything appears marketable.

When Becker earned the Nobel Prize in 1992 "for having extended the domain of microeconomic analysis to a wide range of human behaviour and interaction, including nonmarket behavior" (Nobelprize.org 1992) other researchers had already set out to bring insights and content from the behavioral sciences back into economics. And in 2002, Daniel Kahneman and Vernon Smith received the Nobel Prize for enriching microeconomic analysis with new real world insights.

Since then, behavioral findings have challenged the microeconomic model of an isolated utility maximizer. Biological and cognitive characteristics have been identified as limiting and shaping the rationality of choice. Perhaps even more importantly, the social environment has begun to undermines consumer sovereignty. Rather than deciding in sovereign independence, the purchase decisions of individual consumers are dependent on what others buy, particularly in affluent societies. People define their consumer preferences in relation to what others prefer, and evaluate the utility of their consumption in the light of what others consume. By "choosing the right pond" (Frank 1986) or the wrong pond, whatever the case and the aim may be, consumers determine with whom they compete and compare themselves.

Despite these new findings, there is still no updated model of consumer behavior. Galbraith's diagnosis that "There is no concept of enough or more than enough" (Galbraith 1997, p. XXI) still remains valid 4 years after the financial crises and the discovery of the self-serving mentality of Wall Street and elsewhere. Perhaps there is no conceptual solution to the problem, since the outcome of maximizing consumption depends on other consumers' choices. In this chapter, we look for an empirical solution. Could happiness serve as a rule to evaluate the utility or disutility of consumption? To answer this question we start out with a thick description of consumption patterns in affluent countries that often do not appear to be satisfying or functional. Theories provide an explanation, happiness research may show us ways out. In the conclusion, we speculate on how consumers can benefit from the insights of the current research by discussing institutional implications.

10.2 How We Spend It: Contemporary Purchasing Behavior of a Global Consumption Power

In liberal market economies, consumers decide what to buy. But how much freedom of choice do they enjoy? To what extent do consumers make rational, non-regretted, thoughtful shopping decisions? To learn about consumption behavior in affluent societies, we contrast the purchasing choices of the rich with the choices of those who have limited resources and little room for maneuver, the poor. A trend over time may shed further light on the interdependence of consumer choices.

So, what do consumers who face no immediate budgetary constraints find useful and tasteful? There is no official index of the goods purchased by those in the top tier of wealth distribution (Poterba 2000). But the Forbes magazine compiles a “Cost of Living Extremely Well Index” (CLEWI) of 40 idiosyncratic goods and services which may reflect the needs and wants of the very rich. On the 2011 Index we find, among other things, a Rolls-Royce Phantom for $380,000, an Oyster Sailing Yacht for $3,586,220, a natural Russian sable coat for $240,000, Gucci loafers for $495, a dinner at La Tour D’Argent, Paris for $470 per person, one year of tuition, room, board, and insurance at Harvard University for $52,652, a shotgun by James Purdey & Sons for $197,004 and 45 min with a psychiatrist on Upper East Side, New York for $325. Created in 1976, the CLEWI index has risen significantly faster than the standard consumer price index (CPI), especially since the mid-1990s. Between 2010 and 2011 the CLEWI increased by 4.5 % compared to 3.6 % for the CPI. The rise in demand for very expensive items and brands is unbroken (DeCarlo 2011).

This is also reflected in the sharply increasing prices for art, a luxury good that is in inelastic supply. With $11.57 billion in total global annual revenue, the art market achieved a new record year in 2011. The high-end market has, in particular, grown extraordinarily. “In 2011, the global art auction generated 21 % more than in 2010”, and it gained its highest increase in revenues in China (41.4 %) (Ehrmann 2012, p. 3).

Another indicator is the (re-)emergence of a high-level household service industry. “Catering to the rich—once considered dead-end service work—is now a hot career track.” Official statistics are still missing, but anecdotal evidence shows that butlers, transformed in skill and pay, “are making a comeback” (Frank 2008, p. 19). Yet, even if a “‘good’ butler should command a salary of about $80,000 a year” (Abelson 2011), the goods and services he is able to appreciate during his working day remain unaffordable for him. Are his wants and needs unaffected by his wealthy work environment? More generally, does the rising economic inequality that we have observed in all affluent and emerging countries since the 1980s (Atkinson 2008; Salverda et al. 2011) affect the consumption of the less wealthy and the poor?

If we compare households over time, we find changing patterns. The US provides reliable longitudinal data, and may serve here as a model case. Luxury goods like expensive cars or fees and admissions for entertainment events are rarely an option for poor households. However, it is revealing that the constrained options of the less well-off have shrunk further during recent decades. Ranked by income, the lowest 20 % of all US households spent less for a car in 2010 than in 1984 ($627 instead of $735). In contrast, even after the financial crisis in 2008, the richest 20 % are willing and able to pay 140 % more on average for a car than in 1984. Annual expenses for theatre and concert tickets, or admission fees to sports clubs, are also generally lower now than back in the 1980s. However, the pattern is replicated. The poorest spend less today than in 1984 while the richest households have increased their expenditure by 224 % (Fig. 10.3).

Even consumption of affordable convenience goods diverges progressively among households. What is more, the widening gap is largest for healthy food like fresh fruit, and puts further pressure on the social gradient of obesity in the US today (Ailshire and House 2011) (Fig. 10.4).

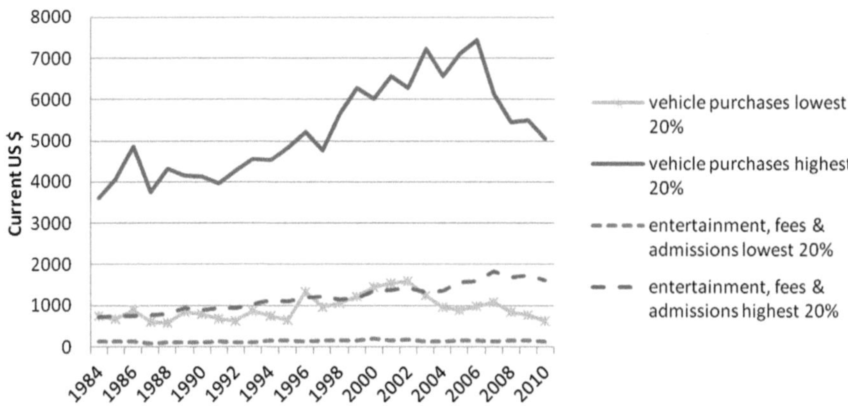

Fig. 10.3 Consumption of luxury goods and leisure activities in US households, by income quintiles. *Source*: Consumer expenditure survey, U.S. Bureau of Labor Statistics 2011

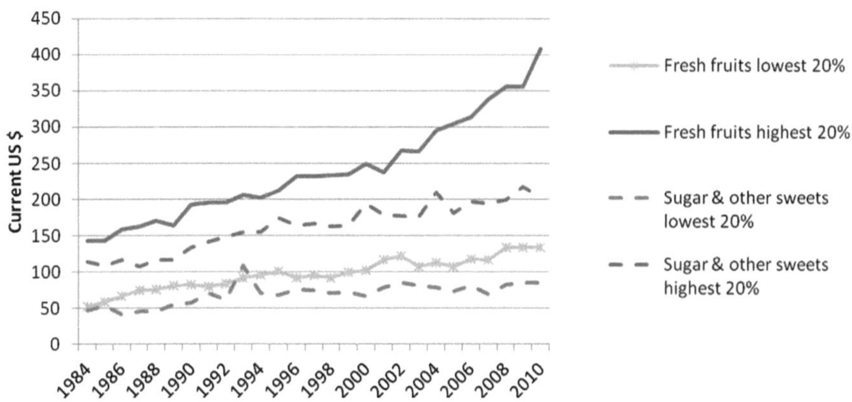

Fig. 10.4 Food consumption in US households by income quintiles. *Source*: Consumer expenditure survey, U.S. Bureau of Labor Statistics 2011

Public health researchers are ringing the alarm bells (Wilkinson and Pickett 2010). Countless studies have shown the detrimental health effects of economic and social inequality. Poverty and low socio-economic status is noxious. It leads to morbid choices. From the Consumer Expenditure Survey we learn that the poorest households in the US today spend more money on tobacco than on medical services. Over recent decades, we have also seen that the consumption of tobacco has increased in low income households while it has declined among high income earners. During the same period, the money spent on medical services stagnated for the poorest and rose sharply for the richest.

Most economists would argue that smoking is an addictive behavior and not a free choice. Still, they preserve a functional explanation and predict: "(…) people lower in socio-economic status with lower life expectancies" have shorter time preferences or "determination" (Schelling 1992, p. 432) and consequently "follow

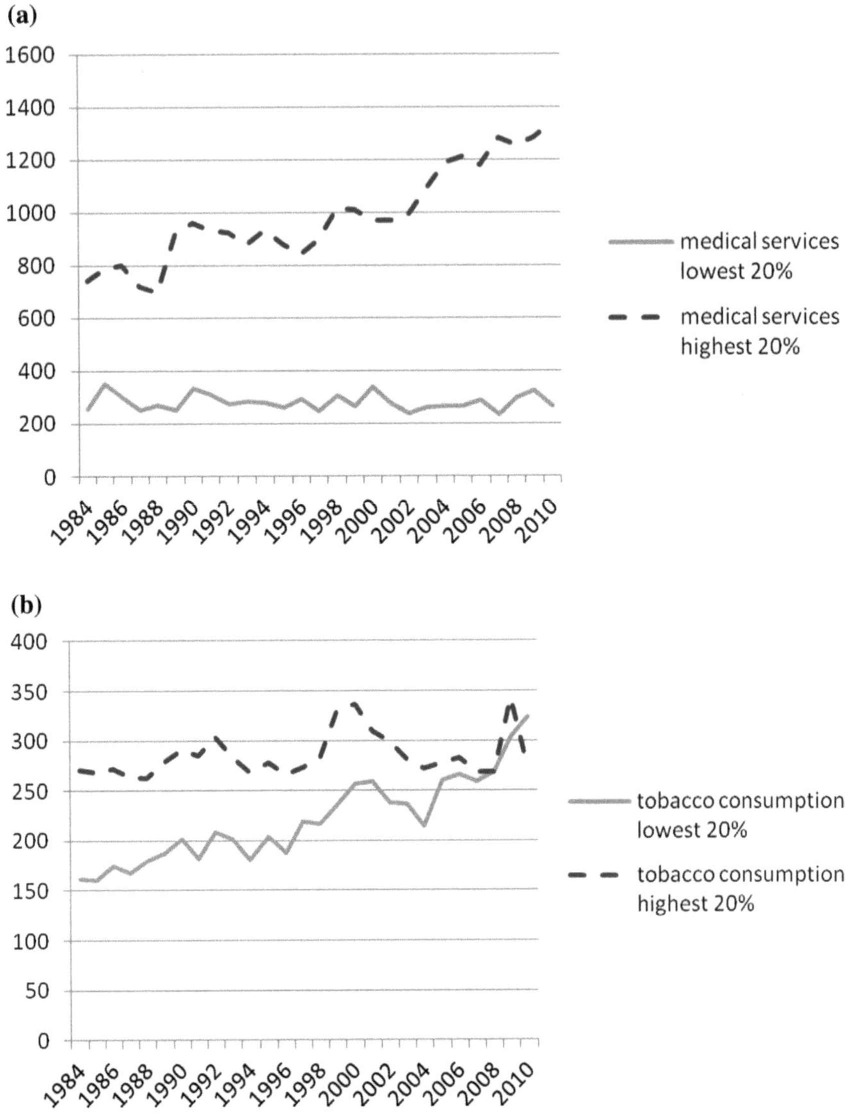

Fig. 10.5 **a** Expenses for medical services in US households by income quintiles. **b** Tobacco consumption in US households by income quintiles. *Source*: Consumer expenditure survey, U.S. Bureau of Labor Statistics 2011

a decade or two behind" the smoking behavior of the upper classes (Schelling 1992, p. 431). But the opposite argument may also be rational. A presumably shorter life expectancy could motivate one to invest in a healthier lifestyle. Regardless of these arguments, the empirical increase in smoking expenditure for the lower income classes contradicts both maximizing expectations (Fig. 10.5).

There are two reasons why the standard economic model fails to predict current consumption patterns. It focuses on maximization aspirations, and it overlooks the impact of the social environment on the consumer. In the following section, we will demonstrate how the happiness paradigm could compensate for both weaknesses. Learning when and how consumers are happy and unhappy gives us a better insight into the logic and dynamics of consumption in affluent countries, and also highlights a way out of an unhealthy and unsustainable spending spree.

10.3 (Un-)Happy Spender: Insights from a New Research Paradigm

Empirical happiness research is mushrooming, but researchers have not been particularly interested in consumers. However, many studies focus on income, and reveal the decisive impact of relative income on individual well-being (Blanchflower and Oswald 2004; Clark et al. 2008; Clark and Oswald 1996; Ferrer-i-Carbonell 2005). When relative income differentials boost happiness independently of absolute income levels, positional goods and services should be in great demand.

Conspicuous Consumption and Other Bad News

Veblen (1899/2001) first described the reputational power of conspicuous consumption in affluent societies. Consumers of unproductive luxuries enjoy the comfort of expensive cars, clothes or champagne, but at the same time they also send out a status message. Duesenberry (1949a, b) expanded this idea to a socially interdependent consumer choice model. He speculates that "above some minimum income level, consumer's satisfaction depends only from the social comparison of one's own consumption with the weighted average consumption of others" (Duesenberry 1949b, p. 178). Consequently, consumers save too little, and invest insufficiently in non-visible goods. They spend unreasonably on positional, attention-grabbing high-price merchandise. In other terms, they objectively reduce their welfare.

The rising levels of income inequality witnessed in the OECD countries and in other parts of the world in recent decades (Atkinson 2008) seemed to fuel the rat race for social status with zero-sum outcome. The US consumption patterns shown above are suggestive. But do they make consumers unhappy? Winkelmann (2012) is one of the few who has tested the effect of conspicuous consumption, measured by density of Ferraris and Porsches, on people's happiness. And he finds that an increase in the number of Ferraris and Porsches per 1,000 of population in Swiss municipalities has a negative effect on income satisfaction, but none on life satisfaction. Also, Hsee et al. (2008) report that inhabitants of large Chinese cities who possess more expensive jewelry are, on average, no happier about their jewelry than the owners of less valuable jewelry living in small Chinese cities.

Positional goods are not inherently valuable and satisfying. Their consumption needs the attention, appreciation and probably sometimes the envy of others. In addition, the quality of positional goods is permanently contested in an affluent environment. Keeping up with the Joneses is stressful. And stress gets under the skin and makes people unhappy. Michael Marmot (2004) first discovered what he called the status syndrome, a causal link between hierarchical rank and health, among Whitehall civil servants. Since then the connection between low social status, high chronic stress and detrimental health effects has been shown to exist in many other human and also non-human populations (Sapolsky 2005). In their bestseller 'The Spirit Level', Wilkinson and Pickett (2010) provide ample evidence of how status inequality, but not income inequality as Goldthorpe (2010) rightly pointed out, may increase the incidence of mental illness, drug abuse, obesity, and life-threatening diseases.

What is more, competition and stress among consumers increase with the number of Joneses and the number of available products. "In an era of global telecommunications and global awareness, only 'the best' assures success in a competition against everybody else" (Schwartz 2005, p. 94). Mass media and the internet broadcast highly selected and very particular lifestyles of people from all over the world, and rarely the lives of Mr. and Mrs. Average next door. Global markets flood our daily lives with endless choice.

Over the past 10 years, for example, the number of products offered in German supermarkets has increased by 130 %, and the number of product variants has increased by as much as 420 %. At the same time, product life cycles have shortened by up to 80 %. Today, every ordinary supermarket offers up to 40,000 products (Schneider 2011). At the same time, a German consulting firm detects room for growth on the luxury market due to "Germans' new affinity for luxury, tourists and **more choices**" (Roland Berger Strategy Consulting 2012).

But maximizing choices does not maximize happiness. On the contrary, empirical evidence shows that extensive choice demotivates people. Iyengar and Lepper (2000) compared consumers and students in limited and extensive-choice situations. In a supermarket they asked one group of customers to evaluate a selection of six different jams while another group had to test 24 different flavors. Even though greater choice initially attracts more consumers, those consumers who had a limited choice could more easily identify the products they liked and were also more likely to purchase those products. Similarly, students felt overwhelmed and unhappy when they had to evaluate 30 different chocolates and not just six. Barry Schwartz speculates that increasing options promotes maximization behavior which, again, lowers people's happiness (Schwartz 2005), since maximization leads to depression, perfectionism, regret, self-blame, and upward comparison (Schwartz et al. 2002).

Then why don't we stop shopping at these supermarkets? There are two answers, a psychological one and a sociological one. First, it seems difficult for anybody to predict correctly how he or she will feel in the future (Gilbert and Wilson 2009). Generally, more choice seems to promise more control and more certainty that our specific preferences and needs will be satisfied. This explains why supermarket costumers are initially attracted by a larger variety of jams. However, to decide in advance when to stop searching, testing, and comparing things is particularly difficult when we have the little information we have about new products. We know more when we compare with others (Gilbert et al. 2009). Sociologists and economists stress the importance of social comparisons which, however, rope consumers into a hedonic treadmill of striving for status and superior lifestyles (Bourdieu 1987; Frank 1986; Veblen 1899/2001).

Be Part of It

Are consumers doomed to be unhappy? Are shoppers least happy in affluent societies? No, happiness research also reveals positive consumption outcomes, if money is wisely spent. Dunn and her colleagues (Dunn et al. 2011) derived several "golden" consumer principles against the background of numerous empirical studies. We discuss the most relevant. The first, "buy experiences instead of things", refers to findings that people are happier with experiential purchases, like an event "that one lives through", than with material goods like cars, clothes or jewelry to which they quickly adapt (Van Boven and Gilovich 2003, p. 1194). We all love to do things, to be engaged, and to get absorbed in activities like creative tasks, hobbies, or vacation trips (Killingsworth and Gilbert 2010). And often we share these flow experiences (Csikszentmihalyi 1999) with others.

Being part of a social community, exchanging and sharing with others is a rewarding, happy experience to which we are rarely habituated. Hence, a second piece of advice for a happy consumer is, "help others instead of yourself". This is backed by a rich empirical literature on altruism and social preferences. In contrast to the standard economic assumption that individuals maximize their egoistic interests, recent experimental research reveals people's satisfaction when they give money away. Harbaugh et al. (2007) traced the happy response of people who both mandatorily and voluntarily transfer money to a charity as far as the level of neural activities in brain areas linked to reward processing. Outside the laboratory, representative samples confirm that pro-social spending correlates with happiness irrespective of people's income (Dunn et al. 2008).

A biological explanation is often put forward. Sexual reproduction calls for sociality. During the course of evolution, assumed to be a functional outcome of selection processes, human beings have become "the most social animal on our planet. Only three other animals (termites, eusocial insects, and naked mole rats) construct social networks as complex as ours, and we are the only one whose complex social networks include unrelated individuals. Many scientists believe that this 'hypersociality' is what caused our brains to triple in size in just 2 million years" (Dunn et al. 2011, p. 117).

Consequently, it is not surprising that satisfaction and happiness evolved in social settings. Sociologists and also economists have specified further conditions under which social norms of selflessness develop and make us happy. Experimental research shows that, in public good games with punishment, a vast majority of subjects behave fairly and cooperatively even if their preferences are initially selfish. In contrast, market games with competition force people to act selfishly even if they would prefer not to do so (Fehr and Schmidt 1999; Güth and Tietz 1990).

These laboratory findings also hold true in the real world (Güth et al. 2007). Families and religious institutions positively sanction altruistic and fair behavior. We know that married and religious people are happier, and they consume more relational goods which "can only be 'possessed' by mutual agreement that they exist, after appropriate joint actions have been taken by a person and non-arbitrary other" (Uhlander 1989, p. 254). Since spending time in and money for voluntary organizations make us feel good (Bruni and Stanca 2008), new forms of collaborative consumption, like product and service sharing, redistributive non-cash markets, and collaborative lifestyles that are growing on the internet should also increase happiness. This consumption promises more control, additional income in kind, and helps to establish new social relationships (Botsman and Rogers 2011).

Another way to overcome near-sighted shopping behavior and credit debts is to engage in postponement strategies. Delayed gratification leads to better and, as our brains reward anticipated pleasures, happier results (Frederick et al. 2003). The psychological advice "pay now and consume later" (Dunn et al. 2011, p. 120) calls the individual to account. Similarly, our human capacity to adapt quickly to changing environments, an asset in many situations, is an obstacle to enjoying consumer goods for a long time. "Buy many small pleasures instead of a few big ones" (Dunn et al. 2011, p. 118) is the suggestion from psychologists to overcome adaptation.

But all these tips may fall short since they overlook the fact that most consumption decisions are not made in a social vacuum. We are affected by other people's consumption styles, particularly in affluent societies. As long as the well-off buy expensive positional goods and services, it is difficult for individuals to drop out of the consumption treadmill. Social scientists call for a nudging architecture, for a larger social framework that applies to everybody. Nudging institutions should control, channel and bind myopic behavior. They should provide incentives and feedback, allow for errors, and ease complex decisions (Thaler and Sunstein 2008). Doesn't that sound familiar?

Perhaps we do not have to reinvent the wheel. Cultural institutions across the globe accommodate the universal pursuit of happiness within a larger social context. Existing frameworks may provide evidence-based insights into the functioning of nudging institutions. They may also help avoid making short-sighted decisions derived directly from lab experiments. Are there cultural role models that can teach us when enough is enough?

10.4 Emerging Collectivistic Role Models?

China

The unparalleled economic growth in China, the size of its markets, the number of its consumers, and its hunger for luxury goods attracts attention. According to Bain Consulting (Bain & Co. 2011), greater China, including Hong Kong and Macau, ranked in 2010 as the third largest luxury goods market, with 17.7 billion € spent. 2 years later China has overtaken Japan and is now the number two market for Cartier, Chanel, Gucci, and other luxury brands. Only US Americans spent more money for high-price merchandise (48.1 billion € in 2010). But China's shoppers, who "account for only 6 % of the world's consumer spending", accounted for "20 % of global sales of luxury goods" (Bain & Co. and Altagamma 2012; Fakes and status 2012, p. 74). Moreover, growth rates are exceptionally high in China (18–22 %) compared to Europe (2–4 %) and the Americas (5–7 %), and sales of luxury goods are expected to remain high.

However, with respect to happiness, China is no growth model. Subjective well-being declined during times of strong economic improvement which pulled millions out of poverty (Wong et al. 2006). This comes as a surprise even compared with previous research. In fact, "luxury fever" is part of the explanation. The hunger for high-price merchandise in China is driven by cultural particularities. The collectivistic orientation of Chinese society emphasizes the fundamental connectedness to each other, and promotes a dominant interdependent self-construal (Markus and Kitayama 1991). Chinese people are likely to affirm their relationships with significant others which, in turn, results in a greater need for social comparison. Moreover, as the desire for social advancement can serve the purpose of greater social acceptance, particularly when strengthened by the key life motive of pursuing "face" (脸) in the Confucian tradition, it is closely related to status thinking and status competition (Chen et al. 2011; Henriksen 2009; Phau and Teah 2009).

Eventually the economic transition from communism to capitalism provided goods and services to express status, or honor, social importance, and appreciation. People who are very concerned about their public self are particularly attracted by conspicuous consumption and material values. A study carried out by Zhang et al. (2011) shows that Chinese who are ranked high on consciousness of face and who want to make a good impression on others tend to be less happy. Other studies (Leary 1996; Schlenker and Leary 1982), using different operationalizations of happiness, confirm this negative relationship. Our own empirical analysis reveals that, in the course of time, the happiness of Chinese people is increasingly determined by relative income and financial satisfaction. This "monetization of happiness" results in more unhappy people, since income inequality has also risen sharply during this period and has turned upward comparison into a frustrating business (Brockmann et al. 2009).

Brazil

Brazil has experienced increasing levels of subjective well-being since 1990,[1] and is the happiest emerging nation in the BRICS world. Based on data from the Gallup World Poll 2009, Brazilians rate their "degree of life satisfaction", measured with an 11-rung Cantril ladder where 10 represents the best possible life, at 8.7 on average, while South Africa and Russia have a value of 5.2, and China and India one of 4.5 (Frayssinet 2011).

Two reasons are often put forward to explain the difference. First, social inequality is falling in Brazil, one of the most unequal countries in the world (Gasparini et al. 2011). "Brazil's Gini dropped from the low 60s in 2000 to somewhere below 57 today—a striking difference given how much relative incomes need to change to effect a 1 Gini point decline or increase, how quickly the change took place, and how unique, compared with the rest of world, it was" (Milanovic 2011, p. 10).

Social support programs (Bolsa Familia), and a broader access to education and skilled work, are held responsible for the improvements. In Brazil, millions of people have climbed into the middle class, but unskilled and less-educated groups have also experienced improvements, while in China, a much more equal society in absolute terms, the best-educated and highest-status groups in urban areas benefited relatively more from the recent economic development than the lower-educated, rural population (Walder 2002; Xing and Li 2012; Zhou 2000).

The second explanation for Brazil's happiness refers back to cultural sources (Graham and Lora 2009; Graham and Sukhtankar 2004; Spector et al. 2004). Not just Brazil but Latin American countries in general rank surprisingly high, given their objective living conditions, on happiness and life satisfaction in surveys (Helliwell et al. 2012).

Social comparison, the underlying mechanism for status and conspicuous consumption, is pervasive in every society. But the process and the effect of social comparisons can differ. Social psychology research (e.g., Festinger et al. 1954; Schachter 1959) has shown that evaluation against a less (more) fortunate other may be ego-enhancing (ego-deflating). Whether social comparison serves a self-enhancement function depends on whether the comparer assimilates or contrasts him/herself to superior or inferior others (Suls et al. 2002). The ego-deflating influence can be associated with the envy effect, where good news for others is bad news for oneself. On the other hand, the self-improvement influence is equivalent to the signal effect, which implies that when individuals evaluate themselves against better-off others, they obtain information about their future prospects and thus generate a positive effect associated with motivation, confidence and inspiration.

Empirical evidence suggests the dominance of a signal effect over an envy effect in social comparisons in Latin American countries where social interactions

---

[1] According to the author's own calculations based on World Values Survey (WVS) 1990–2006.

are also highly esteemed (Ateca-Amestoy et al. 2011). Similarly, studies on developing economies have identified a positive signal effect of relative comparisons (Caporale et al. 2009; Kingdon and Knight 2007; Senik 2004, 2008), while in developed countries the envy effect seems to prevail. So the Brazilian role model may teach us two lessons. First, to improve happiness further in countries with levels of subjective well-being that are already high, social inequality should be rigorously reduced through political measures. And second, a firm social network seems to encourage a positive handling of status competition and consumption. Friends and families help to filter and to interpret other people's lifestyles and consumption patterns as positive information not as personal denigration, and pull people away from status competition. What are the implications of these findings for affluent societies?

10.5 A Political Outlook into a World of Better Proportioned Goods and Services

Back in the rich world, witness conflicting developments. We see that populations are getting more and more obese, while top fashion designers praise zero-size models. Self-restriction is a status symbol in times of excessive supply. However, as long as the individual is left alone with self-restrictive behavior, this may lead to erroneous outcomes. The many pathological cases of anorexia in the fashion industry and in young girls illustrate how difficult it is for a single person to drop out of a treadmill. Other myopic shopping decisions are mentioned throughout the chapter.

But when self-restricting mechanisms are embedded in social or cultural institutions, they create a trusting environment of collectively binding standards that nudge people into overcoming their own short-sighted, wasteful consumer behavior. In this chapter we exploited the new empirical findings in happiness research as a standard for evaluating malconsumption. Many other researchers use the finiteness of our natural resources to define boundaries for sustainable and acceptable consumption (Rockström et al. 2009). The conclusions they draw are similar.

If consumers decide individually in affluent markets, they are easily seduced into spending more and more time and money on positional goods and services. Relative consumption wastes life time and natural resources, and soon makes people unhappy (Benesch et al. 2010; Eaton and Eswaran 2009; Hsee et al. 2009). If they decide in a group, they may invest in collective experiences that absorb time, but not necessarily other resources, and may therefore happily restrain themselves. This is why nutritionists recommend family dinners and not lonely fast food meals.

On a broader scale, happiness researchers, in the same way as environmental researchers, call for political intervention. The Brazilian case could provide evidence that average happiness responds to effective redistributive measures, while rising inequality, as is occurring in China, is making people unhappy. Political redistribution generally means higher taxes. However, against the background of the current happiness findings, there is no accurate optimal taxation schedule which minimizes distortion and inefficiencies. Economists usually strive for such a solution. But instead of using simplistic behavioral assumptions and the toolbox of neo-classical economics solving mathematically abstract maximization problems which "in their general form, are hard to interpret" (Weisbach 2008, p. S296), happiness researchers opt for an evidence-based approach, open to findings from the behavioral and social sciences.

In doing this, Robert Frank (1999) suggests a steeply progressive consumption tax on conspicuous goods and services, to curb status competition. Even though it is not clear if increased prices for Ferraris, Cartier watches, Chanel perfume or large mansions would slow down the rat race for everyone, it would definitely nudge those who have very little and who are most vulnerable in a competitive setting to rethink their consumption strategy. In addition, taxes function as signals. Higher taxes on positional goods and services signal that their consumption is unnecessary and of little value to the community. On the other hand, lower taxes on philanthropic, altruistic donations communicate the highly esteemed status of common goods.

Finally, money raised in tax can be used to narrow the widening gap between the wealthy and the non-wealthy. We do not doubt that economic inequality is probably efficient in some social areas such as the labor markets. But why should economic distinctions penetrate every aspect of life? Why should a lower job rank manifest itself in worse health? Why should children of less fortunate parents suffer from a poorer education? A new tax regime should help people to make better (consumption) decisions not only for themselves but also for society as a whole. It is no surprise that the happiest countries in the world are the Scandinavian nations with high rates of taxation and low social inequality (Oishi et al. 2012).

References

Abelson M (2011) Butler school: in service of those who serve. Bloomberg Businessweek

Ailshire JA, House JS (2011) The unequal burden of weight gain: an intersectional approach to understanding social disparities in BMI trajectories from 1986 to 2001/2002. Soc Forces 90 (2):397–423

Ateca-Amestoy V, Cortés Aguilar A, Moro-Egido AI (2011) Social interactions and subjective well-being: evidence from Latin America. DFAE-II Working Papers, (2011–05), Bilbao

Atkinson AB (2008) The changing distribution of earnings in OECD countries. Oxford University Press, Oxford

Bain & Company (2011) 2011 China luxury market study. Bain point of view. Available at http://www.bain.com/publications/articles/2011-China-luxury-market-study.aspx. Accessed on 28 Dec 2011

Bain & Company, & Fondazione Altagamma (2012) Luxury goods worldwide market study. Spring 2012 update. Available at http://www.altagamma.it/img/sezione3/files/397_975_file.pdf

Benesch C, Frey BS, Stutzer A (2010) TV channels, self-control and happiness. B. E. J Econ Anal Policy 10(1), Article 86
Blanchflower DG, Oswald AJ (2004) Well-being over time in Britain and the USA. J Publ Econ 88(7–8):1359–1386
Botsman R, Rogers R (2011) What's mine is yours. How collaborative consumption is changing the way we live. Harper Collins, London
Bourdieu P (1987) Die feinen Unterschiede. Kritik der gesellschaftlichen Urteilskraft. Suhrkamp, Frankfurt/M
Brockmann H, Delhey J, Welzel C, Hao Y (2009) The China puzzle: falling happiness in a rising economy. J Happiness Stud 10:389–405
Bruni L, Stanca L (2008) Watching alone: relational goods, television and happiness. J Econ Behav Organ 65(3–4):506–528
Caporale GM, Georgellis Y, Tsitsianis N, Yin YP (2009) Income and happiness across Europe: do reference values matter? J Econ Psychol 30(1):42–51
Chen X, Kanbur R, Zhang X (2011) Peer effects, risk pooling, and status seeking. What explains gift spending escalation in rural China? IFPR Discussion paper 01151. International Food Policy Research Institute (IFPRI), Washington
Clark AE, Frijters P, Shields MA (2008) Relative income, happiness, and utility: an explanation for the Easterlin paradox and other puzzles. J Econ Lit 46(1):95–144
Clark AE, Oswald AJ (1996) Satisfaction and comparison income. J Publ Econ 61(3):359–381
Csikszentmihalyi M (1999) If we are so rich, why aren't we happy? Am Psychol 54:821–827
DeCarlo S (2011) Cost of living extremely well index: the price of living large is up Available at: http://www.forbes.com/sites/scottdecarlo/2011/09/26/cost-of-living-extremely-well-index-the-price-of-living-large-is-up/. Accessed on 26 June 2012
Duesenberry JS (1949a) Income, saving, and the theory of consumer behavior. Harvard University Press, Cambridge
Duesenberry JS (1949b [1997]) A reformulation of the theory of saving. In: Goodwin NR, Ackerman F, Kiron D (Eds) The consumer society. Island Press, Washington, DC, pp 176–179
Dunn EW, Aknin L, Norton MI (2008) Spending money on others promotes happiness. Science 319:1687–1688
Dunn EW, Gilbert DT, Wilson TD (2011) If money doesn't make you happy, then you probably aren't spending it right. J Consum Psychol 21(2):115–125
Eaton BC, Eswaran M (2009) Well-being and affluence in the presence of a Veblen good. Econ J 119(539):1088–1104
Ehrmann T (2012) Art market trends 2011. Saint-Romain-au-Mont-d'Or, France. Available at http://www.artmarket.com/trends-2011.html. Accessed on 20 Oct 2012
Fakes and Status in China (2012) The Economist, June 23, p 74
Fehr E, Schmidt KM (1999) A theory of fairness, competition, and cooperation. Q J Econ 114 (3):817–868
Ferrer-i-Carbonell A (2005) Income and well-being: an empirical analysis of the comparison income effect. J Publ Econ 89(5–6):997–1019
Festinger L, Torrey J, Willerman B (1954) Self-evaluation as a function of attraction to the group. Human Relat 7(2):161–174
Frank RH (1986) Choosing the right pond. Human behavior and the quest for status. Oxford University Press, New York
Frank RH (1999) Luxury fever. Weighting the costs of excess. Princeton University Press, Princeton
Frank RH (2008) Richistan. A journey through the 21st century wealth boom and the lives of the new rich. Piatkus Books, London
Frayssinet F (2011) BRAZIL: the 'happiest' emerging nation Available at: http://www.ipsnews.net/2011/07/brazil-the-happiest-emerging-nation/. Accessed on 18 June 2012
Frederick S, Loewenstein G, O'Donoghue T (2003) Time discounting and time preference: a critical review. In: Loewenstein G, Read D, Baumeister RF (eds) Time and decision. Russell Sage Foundation, New York, pp 13–86

Frenzen J, Hirsch PM, Zerrillo PC (1994) Consumption. Preferences, and changing lifestyles. In: Smelser NJ, Swedberg R (eds) The handbook of economic sociology, Princeton University Press, Princeton, pp 403–425
Galbraith JK (1997) Foreword. In: Goodwin NR, Ackerman F, Kiron D (eds) The consumer society, Island Press, New York, pp xxi–xxiii
Gasparini L, Cruces G, Tornarolli L (2011) Recent trends in income inequality in Latin America. Economia 11(2):147–190
Gilbert DT, Killingsworth MA, Eyre RN, Wilson TD (2009) The surprising power of neighborly advice. Science 323(5921):1617–1619
Gilbert DT, Wilson TD (2009) Why the brain talks to itself: sources of error in emotional prediction. Philos Trans R Soc B-Biol Sci 364(1521):1335–1341
Goldthorpe JH (2010) Analysing social inequality: a critique of two recent contributions from economics and epidemiology. Eur Sociol Rev 26(6):731–744
Graham C, Lora E (2009) Paradox and perception: measuring quality of life in Latin America. Brookings Institution, Washington
Graham C, Sukhtankar A (2004) Does economic crisis reduce support for markets and democracy in Latin America? Some evidence from surveys of public opinion and well being. J Latin Am Stud 36:349–377
Güth W, Schmidt C, Sutter M (2007) Bargaining outside the lab—a newspaper experiment of a three-person ultimatum game. Econ J 117(518):449–469
Güth W, Tietz R (1990) Ultimatum bargaining behavior. A survey and comparison of experimental results. J Econ Psychol 11:417–449
Harbaugh WT, Mayr U, Burghart DR (2007) Neural responses to taxation and voluntary giving reveal motives of charitable donations. Science 316:1622–1625
Helliwell J, Layard R, Sachs J (eds) (2012) World happiness report. The Earth Institute, Columbia University, New York
Henriksen M (2009) Luxury fever in China: an analysis of chinese luxury consumption. Doctoral dissertation. Available at: http://books.google.de/books?id=5eY0YAAACAAJ. Accessed on 20 Dec 2012
Hsee CK, Xu F, Tang N (2008) Two recommendations on the pursuit of happiness. J Legal Stud 37:S115–S132
Hsee CK, Yang Y, Li NH, Shen LX (2009) Wealth, warmth, and well-being: whether happiness is relative or absolute depends on whether it is about money, acquisition, or consumption. J Mark Res 46(3):396–409
Iyengar S, Lepper MR (2000) When choice is demotivating: can one desire too much of a good thing? J Pers Soc Psychol 79(6):995–1006
Killingsworth MA, Gilbert D (2010) A wandering mind is an unhappy mind. Science 330:932
Kingdon GG, Knight J (2007) Community, comparisons and subjective well-being in a divided society. J Econ Behav Organ 64(1):69–90
Leary MR (1996) Self-presentation: Impression management and interpersonal behavior. Westview Press, Boulder
Markus HR, Kitayama S (1991) Culture and the self: implications for cognition, emotion, and motivation. Psychol Rev 98(2):224–253
Marmot M (2004) The status syndrome. How your social standing directly affects your health. Times Books, London
Milanovic B (2011) More or less. Income inequality has risen of the past quarter-century instead of falling as expected. Finance and Development, September 2011:6–11
Nobelprize.org (1992) The Sveriges Riksbank prize in economimc sciences in memory of Alfred Nobel 1992. Available at:http://www.nobelprize.org/nobel_prizes/economics/laureates/1992/. Accessed on 31 July 2012
Oishi S, Schimmack U, Diener E (2012) Progressive taxation and the subjective well-being of nations. Psychol Sci 23(1):86–92
Phau I, Teah M (2009) Devil wears (counterfeit) Prada: a study of antecedents and outcomes of attitudes towards counterfeits of luxury brands. J Consum Mark 26(1):15–27

Poterba JM (2000) Stock market wealth and consumption. J Econ Perspect 14(2):99–118
Rockström J, Steffen W, Noone K, Persson Å, Chapin FS III, Lambin EF et al (2009) A safe operating space for humanity. Nature 461:472–475
Roland Berger Strategy Consulting (2012) Roland Berger study on the luxury goods market: Germany has the biggest potential in Europe [Press Release]. Available at: http://www.rolandberger.com/media/press/releases/512-press_archive2012_sc_content/Roland_Berger_study_on_luxury_goods_market.html. Accessed on 29 May 2012
Salverda W, Nolan B, Smeeding TM (eds) (2011) The Oxford handbook of economic inequality. Oxford University Press, Oxford
Samuelson PA (1938) A note of the pure theory of consumer's behaviour. Economica 5(17):61–71
Sapolsky RM (2005) The influence of social hierarchy on primate health. Science 308 (5722):648–652
Schachter S (ed) (1959) The psychology of affiliation. Experimental studies of the sources of gregariousness. Stanford University Press, Stanford
Schelling TC (1992) Addictive drugs: the cigarette experience. Science 255(5043):430–433
Schlenker BR, Leary MR (1982) Audiences' reactions to self-enhancing, self-denigrating, and accurate self-presentations. J Exp Soc Psychol 18:89–104
Schneider RU (2011) Preiskampf in der Bückzone. Neue Züricher Zeitung Folio, November 2011:6
Schwartz B (2005) The paradox of choice. Why more is less. HarperCollins, New York
Schwartz B, Ward A, Monterosso J, Lyubomirsky S, White K, Lehman DR (2002) Maximizing versus satisficing: happiness is a matter of choice. J Pers Soc Psychol 83(5):1178–1197
Senik C (2004) When information dominates comparison—learning from Russian subjective panel data. J Publ Econ 88(9–10):2099–2123
Senik C (2008) Ambition and jealousy: income interactions in the "Old Europe" versus the "New Europe" and the United States. Economica 75(299):495–513
Spector PE, Cooper CL, Poelmans S, Allen TD, O'Driscoll M, Sanchez JI et al (2004) A cross-national comparative study of work-family stressors, working hours, and well-being: China and Latin America versus the Anglo world. Pers Psychol 57(1):119–142
Stigler GJ, Becker GS (1977) De gustibus non est disputandum. Am Econ Rev 67(2):76–90
Suls J, Martin R, Wheeler L (2002) Social comparison: why, with whom, and with what effect? Curr Dir Psychol Sci 11(5):159–163
Thaler RH, Sunstein CR (2008) Nudge. Improving decisions about health, wealth and happiness. Yale University Press, New Haven
Uhlander CJ (1989) Relational goods and participation: incorporating sociality into a theory of rational action. Public Choice 62:253–285
Van Boven L, Gilovich T (2003) To do or to have? That is the question. J Pers Soc Psychol 85 (6):1193–1202
Veblen T (1899[2001]) The theory of the leisure class. Random House, New York
Walder AG (2002) Markets and income inequality in rural China: political advantage in an expanding economy. Am Sociol Rev 67(2):231–253
Weisbach DA (2008) What does happiness research tell us about taxation? [Proceedings Paper]. J Legal Stud 37:S293–S324
Wilkinson R, Pickett K (2010) The spirit level. Why equality is better for everyone, 2nd edn. Penguin Books, London
Winkelmann R (2012) Conspicuous consumption and satisfaction. J Econ Psychol 33(1):183–191
Wong C, Wong K, Mok B (2006) Subjective well-being, societal condition and social policy: the case study of a rich Chinese society. Soc Indic Res 78(3):405–428
Xing CB, Li S (2012) Residual wage inequality in urban China, 1995–2007. China Econ Rev 23 (2):205–222
Zhang X-A, Tian P, Grigoriou N (2011) Gain face, but lose happiness? It depends on how much money you have. Asian J Soc Psychol 14(2):112–125
Zhou XG (2000) Economic transformation and income inequality in urban China: evidence from panel data. Am J Sociol 105(4):1135–1174

Chapter 11
Some Lessons from Happiness Economics for Environmental Sustainability

Heinz Welsch

11.1 Introduction

Until the recent past, the discipline of economics has been severely constrained by the non-observability of two of its most fundamental concepts, preference and utility. This limitation has forced researchers to base their analyses on a set of assumptions or maintained hypotheses that are not able to be tested empirically.

With respect to environmental issues, among the fundamental assumptions usually made in economic analysis are the following:

- In their utility evaluation of market goods and environmental goods, people trade off *absolute levels* of the respective goods against each other.
- When deciding on environmentally relevant behavior, people successfully make individually optimal *utility-maximizing choices.*

While utility, being an abstract concept, is certainly not directly observable, recent research has shown that stated subjective well-being, elicited in surveys, can serve as an empirical proxy for people's experienced utility. Using subjective well-being—or happiness—as a measure of utility, it is possible to test the fundamental issues of utility and choice theory, for instance whether people derive utility from absolute or relative consumption and whether their choices are utility maximizing. The methods and results of happiness research can be used for welfare analysis and benefit–cost studies. The use of happiness data to study economic issues has recently become a burgeoning field in economic literature.

With respect to the principle of maximization, the economics of happiness contributes to what Sent (2004) calls "new" behavioral economics. Whereas "old" behavioral economics questions whether economic behavior involves utility maximization rather than, say, rule-based behavior, new behavioral economics

H. Welsch (✉)
Department of Economics, Carl von Ossietzky Universität, Oldenburg, Germany
e-mail: welsch@uni-oldenburg.de

H. Brockmann and J. Delhey (eds.), *Human Happiness and the Pursuit of Maximization*, Happiness Studies Book Series, DOI: 10.1007/978-94-007-6609-9_11,

takes the mainstream model of utility maximization as a benchmark and studies deviations from that model. In new behavioral economics, the issue is not whether people *strive* to maximize utility, but whether or not they *succeed* in doing so.

In happiness economics, the arguments of the utility function are not merely economic variables (levels of income and consumption), but also leisure, social life, environmental amenities and other factors of well-being, and an important question in happiness economics is whether economic choices lead to an optimal (that is, utility maximizing) balance of the various factors of utility.

Against this background, the purpose of this chapter is to show that happiness research has considerable potential for the analysis of environmental sustainability. Being an empirical indicator of utility, happiness data permit the testing of the assumptions usually made in environmental economics and the study of environmental issues independent of the validity of these assumptions. In relation to the assumptions mentioned above, this chapter shows that:

- The importance of relative consumption levels found in happiness research (as opposed to absolute consumption levels) implies more ambitious targets for socially optimal environmental policy.
- Happiness research suggests that people's environment-friendly consumption may be lower than is individually optimal (that is, less than utility maximizing).

The chapter starts by discussing happiness as an empirically applicable approximation to the traditional economic notions of utility and preference. It then shows how the economic analysis of happiness has uncovered previously neglected channels through which the current styles of economic behavior affect environmental integrity and sustainability. Based on these results, the chapter concludes with a brief discussion of the ways in which a better understanding of the factors and mechanisms underlying happiness may contribute to forming a more environmentally sustainable lifestyle.

11.2 Happiness and the Maximization of Utility

Happiness and Utility

"Happiness" denotes a measure of an individual's evaluation of her overall quality of life (Veenhoven 1997). The term is usually used interchangeably with "life satisfaction." The umbrella term that encompasses both concepts is "subjective well-being." Data on subjective well-being have been introduced into economics as an empirical approximation to a notion of utility labeled "experienced utility" by Kahneman et al. (1997). Experienced utility is the *ex post* hedonic quality (satisfaction) associated with an act of choice, in contrast to decision utility, which is the *ex ante* expectation of experienced utility. Experienced utility is the utility concept used by the classics, and its application in contemporary economics may be referred to as neo-utilitarianism.

Data on individuals' subjective well-being are elicited in large-scale surveys and are being used in the growing literature in economics. Some of the relevant surveys refer to single countries, such as the *General Social Surveys* in the U.S. or the *German Socio-Economic Panel*. Others, like the *Eurobarometer Surveys* or the *World Values Surveys*, use a common format for eliciting subjective well-being for several countries.

The questions pertaining to subjective well-being may refer to "happiness" or to "life satisfaction," and the categories may be purely verbal or may combine verbal with numerical features. For instance, the *General Social Surveys* use a three-point verbal happiness scale, which asks the question: "Taken all together, how would you say things are these days—would you say that you are very happy, pretty happy, or not too happy?" The *Eurobarometer Surveys* use a four-point verbal life satisfaction scale, employing the question: "On the whole, are you very satisfied, fairly satisfied, not very satisfied, or not at all satisfied with the life you lead?" In the *World Values Surveys*, people are offered a scale from 1 (dissatisfied) to 10 (satisfied) to respond to the question: "All things considered, how satisfied are you with your life as a whole these days?"

A precondition for using happiness or life satisfaction data as a proxy for utility is that they satisfy the appropriate quality requirements. In particular, a basic condition is that the data are at least ordinal in character and satisfy the conventional quality standards. Whether these conditions are satisfied has been widely assessed in decades of validation research (see, e.g., Frey and Stutzer 2002b, for references). In these studies, measures of subjective well-being are generally found to have a sufficient degree of internal consistency, validity and reliability as well as a high degree of stability over time (Diener et al. 1999). Different measures of subjective well-being—especially measures of happiness and of life satisfaction—correlate well with each other and, according to factor analyses, represent a single unitary construct. Happiness responses are correlated with physical reactions that can be thought of as describing true, internal happiness: people who report that they are happy tend to smile more and show lower levels of stress responses (heart rate, blood pressure), and they are less likely to commit suicide. Overall, measures of reported subjective well-being can be viewed as valid and reliable empirical approximations to individual utility.

In addition to their validity and reliability properties, happiness data not only need to be (at least) ordinal, but they also need to be comparable in the sense that people share a common opinion of what happiness is. This assumption relies on supporting evidence from psychology (Diener et al. 1999). One indication is that individuals are able to recognize and predict the happiness level of others. Happy people are usually rated by others as happy. While comparability is a stronger assumption than economists frequently need to make, it may be less problematic at a practical level than suggested by theorists (Kahneman 1999).

Arguments of the Happiness/Utility Function

Research on happiness has identified a number of personal, demographic and socio-economic covariates of happiness that explain the observed happiness patterns. Important personal and demographic characteristics that affect happiness are health, age, sex, marital status, the size and structure of the household, the education level and the degree of urbanization.

The socio-economic determinants of happiness can be classified into factors at the micro (individual) level and factors at the macro (societal) level. At the micro level, personal income (or household income) and employment status are important. With respect to personal income, a robust finding is that income does increase happiness, and that both absolute income and relative income play a role (Clark et al. 2008). Regarding the employment status, being unemployed shows a strong negative association with happiness; this is true even when controlling for income. Personal unemployment is the strongest micro-level factor for unhappiness (Frey and Stutzer 2002b).

Important factors at the societal level are macroeconomic conditions (unemployment rate, inflation rate, growth rate), institutional conditions (political freedom, democracy, the rule of law), public evils (terrorism, civil war, corruption) and environmental amenities. The unemployment rate and the inflation rate affect happiness negatively (Di Tella et al. 2001), whereas the growth rate affects happiness positively (Welsch 2007b). Good institutional quality yields greater happiness (Frey and Stutzer 2002a). Terrorism, civil war and corruption have sizeable negative effects on happiness (Frey et al. 2009; Welsch 2008a, b, respectively).

Most importantly in the present context, environmental amenities also affect happiness. The environmental factors addressed so far in the available studies cover a considerable range of environmental problems and several forms of environment-related extreme events. They comprise air pollution (Ferreira and Moro 2010; Luechinger 2009; MacKerron and Mourato 2009; Menz and Welsch 2010, 2011; Welsch 2002, 2006, 2007a), water pollution (Israel and Levinson 2003), airport noise (van Praag and Baarsma 2005), climate parameters (Rehdanz and Maddison 2005), flood events (Luechinger and Raschky 2009) and drought events (Carroll et al. 2009). All of these studies found that environmental amenities affect happiness positively, whereas environment-related extreme events affect happiness negatively.

Coherence of the Utility Function

As mentioned above, the behavioral economics literature distinguishes conceptually between experienced utility and decision utility and uses happiness data as an empirical approximation of the former. Against this conceptual framework, an important *empirical* issue is whether or not the decision utility function and the experienced utility function coincide: in other words, whether people know and correctly apply their experienced utility function when making economic choices. If not, this would constitute a radical departure from the standard economic assumption that people correctly evaluate their own preferences and make utility-maximizing choices.

While the latter would not constitute a major problem if the decision errors were small in size and random in character, the available evidence suggests that such errors may be systematic. In particular, as will be discussed below, there may be systematic decision errors to the disadvantage of non-material choices, such as pro-environmental consumption. This implies that people could increase their utility by making less materialistic choices.

11.3 Consumption Externalities and Market Failure

Relative Consumption Effects

A standard assumption in economic theory, which will prove important for optimal environmental regulation, is that people evaluate consumption goods without regard to other people's consumption and their own past consumption behavior. In contrast to this assumption, behavioral economics in general and the happiness literature in particular have found that people evaluate consumption relative to certain benchmarks, namely other people's consumption (social comparison) and their own consumption in the past (adaptation); see Clark et al. (2008) for a survey.

Based on such evidence, the theoretical literature has investigated the implications of relative concerns for public policy. In particular, models have been analyzed in which social comparison implies that a person's evaluation of her consumption level is negatively affected by other people's consumption level: in other words, that other people's consumption yields a negative externality. In a similar fashion, adaptation implies that a person's evaluation of her current consumption level is negatively affected by her own past consumption, such that the past consumption yields a negative "internality." Importantly, adaptation to consumption levels is largely unforeseen in the individual's decision making, due to a failure of affective forecasting (Gilbert et al. 1998; Loewenstein et al. 2003; Wilson and Gilbert 2003).

If, in such circumstances, people are faced with the choice between work time—which provides the resources for consumption—and leisure time, they will overwork (overconsume) relative to the social optimum. That is, everybody would be better off if they worked (consumed) less, thus imposing fewer consumption externalities on each other.

In economics jargon, this is an instance of market failure, which calls for correction by public policy, for instance by means of a consumption or income tax (Layard 2006).

Implications for Environmental Regulation

The standard economic rationale for environmental policy relies on the concept of environmental externalities. This means that a person's (or firm's) economic activities leads to a deterioration in environmental quality that affects the well-being (utility) of other persons. In their choice of economic activity levels, people will (rationally) disregard these effects on other persons' well-being. The outcome from such behavior is that in an unregulated economy the level of economic activity is too high and the level of environmental quality is too low compared with what is socially optimal. There is thus market failure due to unregulated environmental externalities, which calls for intervention by public policy.

This kind of standard analysis focuses on physical environmental externalities from economic activities, say consumption. If, in addition to those physical externalities, psychological externalities exist, as discussed in the preceding sub-section, then unregulated individual utility maximization will imply levels of economic activity that are too high due to both types of externality *combined* and levels of environmental quality that are too low (relative to the social optimum). In other words, the optimal level of environmental quality that arises when both types of externality are jointly present is higher than in the standard model of environmental policy (which disregards psychological consumption externalities).

The overall conclusion from this discussion is that negative consumption externalities and unanticipated adaptation to consumption levels—both of which have been identified in recent happiness research—may imply distortions away from the optimal environmental quality *in addition* to the familiar distortions stemming from environmental externalities. Optimal environmental regulation may thus be stricter than suggested by the standard model of environmental policy analysis.

11.4 Decision Error and Overconsumption

Social Optimality and Individual Optimality

The discussion of the preceding section disregarded the possibility that people voluntarily engage in pro-environmental activities. In that framework, environmental externalities and consumption externalities imply that the individually optimal outcome of (unregulated) consumer choice fails to be *socially* optimal (market failure).

If voluntary pro-environmental activities are present, an additional issue arises: will the chosen level of pro-environmental activity be *individually* optimal (utility maximizing) or different (*ex post*) from the individual utility maximum—due to a divergence between decision utility and experienced utility? As detailed below, the results obtained from happiness economics suggest that people choose levels of pro-environmental activity that are less than optimal even in terms of their own individual utility evaluation.

Utility Misprediction and its Consequences

As discussed above, people adapt to the consumption levels attained but fail to anticipate this adaptation correctly when taking consumption decisions, due to a failure in affective forecasting (Gilbert et al. 1998; Loewenstein et al. 2003; Wilson and Gilbert 2003). However, such adaptation does not seem to apply to all sorts of activities and outcomes alike. Especially, people do not seem to adapt their utility evaluation in the case of outcomes that relate to so-called intrinsic motivation, as opposed to extrinsic motivation.[1] In the case of intrinsic motivation, utility derives from an internal reward as a direct result of a particular activity or choice. At a fundamental level, intrinsic motivation has been linked to a need for relatedness, competence or autonomy (Deci and Ryan 2000). In the case of extrinsic motivation, choice is instrumental to an external goal, such as acquisition, possession, status or prestige.

A major example of a lack of adaptation to outcomes that relate to intrinsic motivation is unemployment. Unemployment affects the need for relatedness, competence and autonomy and has consistently been found to have large and *persistent* negative effects on subjective well-being (e.g., Clark et al. 2008). By contrast, changes in income (or consumption) largely relate to extrinsic motivation and are subject to a considerable degree of adaptation, such that the effects tend to be transitory.

Since the failure of affective forecasting results from the failure to anticipate hedonic adaptation, and since hedonic adaptation is more important for some categories of outcomes than for others, it follows that some sorts of outcomes are more liable to inaccurate utility forecasting than others. This asymmetry in the accuracy of utility forecasting is an origin of distorted, non-utility-maximizing choices. In particular, it implies that a choice is distorted towards consumption (or income) relative to activities that serve less material goals. Moreover, this distortion is likely to be larger the more weight people place on material relative to non-material outcomes (Stutzer and Frey 2007).

Frey and Stutzer (2004) addressed commuting as an example of an extrinsically motivated activity, whereas Meier and Stutzer (2008) studied volunteering as an example of an intrinsically motivated activity. These studies found negative (net) marginal utility from commuting and positive (net) marginal utility from volunteering, respectively. That is, people could raise their utility by commuting less and by volunteering more. Thus, these findings are inconsistent with the idea of

[1] For these concepts, see Maslow (1968), Rogers (1961), Kasser and Ryan (1996) and Frey and Stutzer (2004).

individually optimal choices. In contrast, they are consistent with an *ex ante* overvaluation of extrinsically motivated activities (serving income acquisition) relative to intrinsically motivated activities (related to human relationships).

Environment-Friendly Consumption and Optimal Consumer Choices

To the extent that pro-environmental consumption is driven by altruism and a "warm glow" (Andreoni 1990), it may be viewed as intrinsically motivated behavior. In the light of the preceding subsection, the question then arises of whether people's choice between the pure level of consumption and the environmental friendliness of consumption is individually optimal or distorted away from environmental friendliness.

This question has been studied by Welsch and Kühling (2010). They considered a utility function in the level of consumption (quantity) and its environmental friendliness (quality). By assuming that the unit cost of consumption increases in environmental friendliness, the attainable quantity is linked to the quality via a budget constraint: due to a limited income, a trade-off exists between the quantity that can be consumed and the environmental friendliness of consumption. In such a framework, a utility-maximizing choice would imply that the marginal utility from consuming in a more environmentally friendly way net of the quantity foregone (due to the budget constraint) is zero.

To be more precise, the formal model can be stated as follows: suppose that individual i possesses a utility function in the level of consumption (quantity), x_i, and its environmental friendliness (quality), q_i:

$$u_i = U(x_i, q_i), \tag{11.1}$$

which is assumed to have standard properties (i.e., it is increasing and strictly concave in both arguments).

The budget constraint that links the attainable quantity to the environmental friendliness (quality) is:

$$x_i = G(q_i, y_i), \quad \partial G/\partial q_i < 0, \quad \partial G/\partial y_i > 0, \tag{11.2}$$

where y_i denotes income. Here, $-\partial G/\partial q_i$ is the marginal cost, in terms of the quantity foregone, of consuming in a more environmentally friendly manner.

Substituting (11.2) into (11.1) yields a semi-reduced utility function

$$u_i = U(G(q_i, y_i), q_i) = V(q_i, y_i), \tag{11.3}$$

whose maximum with respect to q_i satisfies the condition that the individual's marginal utility from consuming in an environmentally friendly way net of the marginal cost is zero:

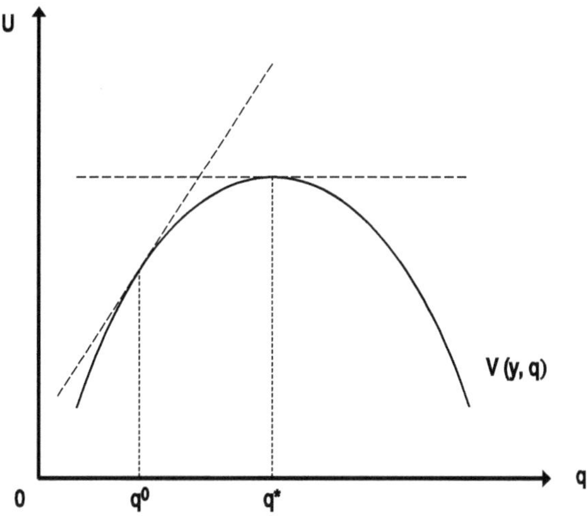

Fig. 11.1 Reduced-form utility function

$$\frac{\partial V(q_i, y_i)}{\partial q_i} = 0. \tag{11.4}$$

This is illustrated in Fig. 11.1: the function *V*(.) is hump-shaped in q_i and has its maximum at the point at which its slope is zero. A positive slope indicates that the level of q_i is less than utility maximizing, whereas a negative slope indicates that the level of q_i is greater than utility maximizing.

The semi-reduced utility function is useful because it permits the utility maximization assumption to be checked with the available data. Specifically, the empirical validity of condition (11.4) can be checked by means of the following empirical specification corresponding to (11.3):

$$H_i = \alpha + \beta \cdot q_i + \gamma \cdot \ln y_i + \delta \cdot controls_i + \varepsilon_i, \tag{11.5}$$

where H_i denotes the happiness (life satisfaction) of individual *i* and ε_i denotes the error term. The net marginal utility from environmental friendliness is β, which should not be significantly different from zero if the choice between consumption level and environmental friendliness is (individually) optimal. Against the background of Fig. 11.1, a positive estimate of this coefficient indicates that environmental friendliness is less than optimal.

By estimating such a model with data on self-reported life satisfaction and on the purchase of environmentally friendly goods from the *World Value Surveys*, Welsch and Kühling (2010) found that the net marginal utility of environmental friendliness is significantly positive. This refutes the hypothesis that the choice of environmentally friendly goods is individually optimal and suggests that the observed environmental friendliness of consumption is less than is individually rational. The quantity consumed, on the other hand, seems to be too large relative to the utility maximum.

It should be noted that such an investigation of the optimality of consumer choice would not be possible without the availability of an empirical measure of utility, as provided by data on happiness: happiness data permit the measurement of economic choices and their utility consequences independently of each other. This allows for a test of the rational choice hypothesis of consumer behavior.

Preference Spillovers and Social Learning

In a case study for Germany, Welsch and Kühling (2011) found confirmation of the results just discussed. The survey data used in that study contain information not just on the pro-environmental behavior of the respondents, but also on the history of this behavior (for how long) and on the corresponding behavior of the respondents' reference persons (friends, neighbors, relatives).

Using these unique data, it was found that the decision errors with respect to pro-environmental consumption are a decreasing function of the experience people have in consuming environmentally friendly goods. In addition, the decision errors are smaller the more intensively people's reference persons consume environmentally friendly goods.

Table 11.1 shows examples of those estimation results with several empirical indicators of pro-environmental consumption (*pec*). These indicators refer to purchases of organic food, low-energy light bulbs, energy-saving household appliances and "green" electricity and the installation of solar panels. The pro-environmental purchase behaviors are coded as never = 1, occasionally = 2, often = 3 or always = 4 or as never/occasionally = 0 or often/always = 1 (see the explanation in the table notes).

The coefficients for *pec* are significantly positive for all the indicators, which shows that people could become happier by consuming in a more environmentally friendly way. However, as the negative interaction terms with the corresponding behaviors of people's friends, neighbors and relatives ("peers") show, underconsumption of environment-friendly goods is smaller as peers display more environmentally friendly behavior.

These results suggest that there may be a process of social learning regarding how to consume in an environmentally friendly fashion, a process that diminishes the gap between people's actual level of pro-environmental behavior and the level that would be utility maximizing.

11.5 Conclusions

Survey data on happiness are increasingly being used in economics. Such data are useful because happiness (or life satisfaction) provides an empirical proxy for utility. The use of such data permits the testing of fundamental assumptions of

Table 11.1 The impact of pro-environmental consumption (Pec) on life satisfaction.

	Pec = consumption1		Pec = consumption2		Pec = consumption3	
	OPM	OLS	OPM	OLS	OPM	OLS
Income	0.107^{a} (4.06)	0.112^{a} (4.14)	0.092^{a} (3.47)	0.099^{a} (3.54)	0.102^{a} (3.82)	0.108^{a} (3.85)
Pec	0.216^{b} (2.37)	0.250^{b} (2.13)	0.475^{a} (3.05)	0.544^{a} (2.58)	0.429^{a} (3.29)	0.489^{a} (2.83)
Pec^{c}Peers	−0.074^{b} (2.08)	−0.083^{c} (1.83)	−0.190^{a} (2.64)	−0.218^{b} (2.13)	−0.154^{a} (3.08)	−0.170^{b} (2.35)
Peers	0.646^{b} (2.05)	0.697^{c} (1.67)	0.603^{b} (2.45)	0.664^{c} (1.85)	0.639^{a} (2.93)	0.670^{b} (2.08)
N	326	326	333	333	321	321
(Pseudo-)R^2	0.104	0.296	0.106	0.302	0.109	0.307

Dependent variable: life satisfaction (*LS*). Method: ordered probit model (*OPM*), ordinary least squares (*OLS*). Heteroskedasticity-robust z-statistics (t-statistics) in parentheses

^{a,b,c} denote significance at the 1, 5 and 10 % level

Respectively. *pec* = pro-environmental consumption

Consumption1 = food (1–4) + lighting (1–4) + appliances (1–4) = 3, 4, …, 12

Consumption2 = 1 + lighting (0–1) + appliances (0–1) + solar (0–1) + electricity (0–1) = 1, 2, … 5

Consumption3 = 1 + food (0–1) + lighting (0–1) + appliances (0–1) + solar (0–1) + electricity (0–1) = 1, 2, … 6

The regressions include socio-demographic characteristics and environment-related attitudes. *Source* Welsch and Kühling (2011)

conventional utility and choice theory and the use of the results for welfare analysis and benefit–cost studies. As discussed in this chapter, the economic research on happiness has considerable implications for the study of environmental sustainability.

A major finding from happiness research is that people evaluate their consumption levels relative to other people's consumption and to their own consumption in the past. This gives rise to negative consumption externalities and to adaptation, the latter being largely unforeseen in the process of consumer choice. Negative consumption externalities and unanticipated adaptation to consumption levels imply distortions away from the optimal environmental quality *in addition* to the familiar distortions stemming from environmental externalities. Socially optimal environmental quality targets may therefore be more ambitious than implied by the standard model of environmental economics.

Happiness data permit the measurement of people's choices and the utility they derive from them independently from each other. While standard microeconomics takes utility maximization as a non-testable axiom, happiness data allow for a test of this hypothesis. The findings from happiness research suggest that consumer choice is not utility maximizing and systematically distorted towards extrinsically motivated options and away from intrinsically motivated options. Initial evidence suggests that such a bias may apply to environment-friendly consumption, being one form of intrinsically motivated behavior. This implies that pro-environmental consumption is not only sub-optimal with respect to the social optimum, but also relative to individually rational behavior.

There is also some evidence that social learning may help reduce the decision errors that lead to sub-optimal levels of pro-environmental behavior. It may, however, also be the case that the importance of relative consumption levels in the happiness equation increases over time, due to the spread of consumption-oriented role models through the media, say. This may give rise to an acceleration of the consumption "arms race," the implications of which for environmental sustainability are anything but desirable. The use of happiness data may provide an avenue towards clarifying the prospects of these potentially opposing developments.

This chapter has evolved around the subject of "economic maximization," understood as a methodological primitive: people strive to attain the maximum level of well-being. Happiness research has not refuted (and probably cannot refute) this axiom. The question, however, is which factors—economic and non-economic—contribute to the goal of maximum well-being. Happiness research has shown that maximization solely of the economic factors certainly will not.

References

Andreoni J (1990) Impure altruism and donations to public goods: a theory of warm-glow giving. Econ J 100:464–477

Carroll N, Frijters P, Shields MA (2009) Quantifying the costs of drought: new evidence from life satisfaction data. J Population Econ 22:445–461

Clark AE, Frijters P, Shields MA (2008) Relative income, happiness and utility: an explanation for the Easterlin paradox and other puzzles. J Econ Lit 46:95–144
Deci EL, Ryan RM (2000) The “what” and “why” of goal pursuits: human needs and the self-determination of behavior. Psychol Inq 11:227–268
Diener E, Suh EM, Lucas RE, Smith HL (1999) Subjective well-Being: three decades of progress. Psychol Bull 125:276–302
Di Tella R, MacCulloch R, Oswald A (2001) Preferences over inflation and unemployment: evidence from surveys of happiness. Am Econ Rev 91:335–341
Ferreira S, Moro M (2010) On the use of subjective well-being data for environmental valuation. Environ Resour Econ 46:249–273
Frey BS, Stutzer A (2002a) Happiness and economics: how the economy and institutions affect well-being. Princeton University Press, Princeton and Oxford
Frey BS, Stutzer A (2002b) What can economists learn from happiness research? J Econ Lit 40:402–435
Frey BS, Stutzer A (2004) Economic consequences of mispredicting utility, Working Paper No. 218. Institute for Empirical Research in Economics, University of Zurich
Frey BS, Luechinger S, Stutzer A (2009) The life satisfaction approach to valuing public goods: the case of terrorism. Public Choice 138:317–345
Gilbert DT, Pinel EC, Wilson TD, Blumberg SJ, Wheatley TP (1998) Immune neglect: a source of durability bias in affective forecasting. J Pers Soc Psychol 75:617–638
Israel D, Levinson A (2003) Examining the relationship between household satisfaction and pollution. Manuscript, Indiana State University, Indiana
Kahneman D (1999) Objective happiness. In: Kahneman D, Diener E, Schwarz N (eds) Well being: the foundations of hedonic psychology. Russell Sage Foundation, New York, pp 3–25
Kahneman D, Wakker PP, Sarin R (1997) Back to Bentham? Explorations of experienced utility. Quart J Econ 112:375–405
Kasser T, Ryan RM (1996) Further examining the American dream: differential correlates of intrinsic and extrinsic goals. Pers Soc Psychol Bull 22:280–287
Layard R (2006) Happiness and public policy: a challenge to the profession. Econ J 116:C24–C33
Loewenstein G, O’Donoghue T, Rabin M (2003) Projection bias in predicting future utility. Quart J Econ 118:1209–1248
Luechinger S (2009) Valuing air quality using the life satisfaction approach. Econ J 119:482–515
Luechinger S, Raschky PA (2009) Valuing flood disasters using the life satisfaction approach. J Public Econ 93:620–633
MacKerron G, Mourato S (2009) Life satisfaction and air quality in London. Ecol Econ 68:1441–1453
Maslow A (1968) Towards a psychology of being, 2nd edn. Van Nostrand, New York
Meier S, Stutzer A (2008) Is volunteering rewarding in itself? Economica 75:39–59
Menz T, Welsch H (2010) Population aging and environmental preferences in OECD countries: the case of air pollution. Ecol Econ 69:2582–2589
Menz T, Welsch H (2011) Life-cycle and cohort effects in the valuation of air pollution: evidence from subjective well-being data. Land Economics 88:300–325
Rehdanz K, Maddison D (2005) Climate and happiness. Ecol Econ 52:111–125
Rogers C (1961) On becoming a person. Houghton Mifflin, Boston
Sent EM (2004) Behavioral economics: how psychology made its (limited) way back into economics. Hist Polit Econ 36:735–760
Stutzer A, Frey BS (2007) What happiness research can tell us about self-control problems and utility misprediction. In: Frey BS, Stutzer A (eds) Economics and psychology: a promising new cross-disciplinary field. MIT Press, Cambridge, pp 169–195
van Praag B, Baarsma B (2005) Using happiness surveys to value intangibles: the case of airport noise. Econ J 52:111–125
Veenhoven R (1997) Advances in understanding happiness. Revue Quebecoise de Psychologie 18:29–74

Welsch H (2002) Preferences over prosperity and pollution: environmental valuation based on happiness surveys. Kyklos 55:473–494
Welsch H (2006) Environment and happiness: valuation of air pollution using life satisfaction data. Ecol Econ 58:801–813
Welsch H (2007a) Environmental welfare analysis: a life satisfaction approach. Ecol Econ 62:541–551
Welsch H (2007b) Macroeconomics and life satisfaction: revisiting the "misery index". J Appl Econ 10:237–251
Welsch H (2008a) The social costs of civil conflict: evidence from surveys of happiness. Kyklos 61:320–340
Welsch H (2008b) The welfare costs of corruption. Appl Econ 40:1839–1849
Welsch H, Kühling J (2010) Pro-environmental behavior and rational consumer choice: evidence from surveys of life satisfaction. J Econ Psychol 31:405–420
Welsch H, Kühling J (2011) Are pro-environmental consumption choices utility maximizing? The role of learning. Ecol Econ 72:75–87
Wilson TD, Gilbert DT (2003) Affective forecasting. In: Zanna MP (ed) Advances in experimental social psychology, vol 35. Elsevier, New York, pp 345–411

Chapter 12
Public Policy and Human Happiness: The Welfare State and the Market as Agents of Well-Being

Robert Davidson, Alexander C. Pacek and Benjamin Radcliff

12.1 Introduction

Market principles are the basis of economic and, increasingly, social organization in the modern world. While the dominance of the market has largely ceased to be contested, disagreement over the scope and power of the market has long been the major political axis of conflict in contemporary politics. Simply put, the issue is the *degree* to which society should be subordinated to the self-regulating control of the market. The Right, as manifested most explicitly in what has been called by its detractors "market fundamentalism" (Soros 1998), argues for the maximum of such subordination. The conventional Left, in the form of Labor and Social Democratic parties and their associated labor movements, argues for less power to the market, through a program of supplementing market outcomes with political interventions that seek to make the democratic state—which is to say, the democratic process itself—the ultimate guarantor of citizen well-being.

Recent decades have witnessed the emergence of a social scientific research program aimed at understanding the empirical consequences of these two alternative philosophical approaches to crafting a nation's public policy regime. We thus find contemporary literatures devoted to determining whether such political "intrusions" into the market achieve their objectives of reducing poverty and inequality, whether they have unintended (and generally deleterious) consequences

R. Davidson (✉)
Amsterdam Research Center for Gender and Sexuality, University of Amsterdam, Amsterdam, The Netherlands
e-mail: R.J.Davidson@uva.nl

A. C. Pacek
Department of Political Science, Texas A&M University, College Station, TX, USA
e-mail: e339ap@polisci.tamu.edu

B. Radcliff
Department of Political Science, University of Notre Dame, Notre Dame, IN, USA
e-mail: radcliff.1@nd.edu

H. Brockmann and J. Delhey (eds.), *Human Happiness and the Pursuit of Maximization*, Happiness Studies Book Series, DOI: 10.1007/978-94-007-6609-9_12,

for economic growth, whether they affect rates of social deviancy (such as violent crime), and whether they are complicit in promoting a variety of social pathologies, such as "cultures of dependency," higher divorce rates, and so on.

Of course, in the end we concern ourselves with all of these issues, from divorce to economic growth, because of their presumed impacts on the quality of human life. That is, we presume that such outcomes, through both direct and indirect causal mechanisms, ultimately make people more or less satisfied with their lives. In this chapter, we attempt to take this fact seriously by focusing our attention not on any of the individual or particular effects of market interventions that in turn are thought to have some effect on quality of life, but instead on quality of life itself. Put another way, rather than considering the presumptive effects of political restraints on the market on intervening variables that are further presumed to have a potential influence on satisfaction with life, we examine directly the connection between political control of the market and life satisfaction. In particular, we ask whether cross-national differences in market dominance affect the degree to which citizens lead lives that they themselves regard as positive and rewarding.

This is now possible, given the development of a sophisticated literature devoted to studying life satisfaction. With the refinement of the tools necessary to measure with reasonable reliability and validity how people subjectively evaluate the quality of their lives, we are now capable of testing theoretically derived hypotheses about the observable factors that tend to affect subjective well-being (Radcliff 2013). In sum, we are capable of measuring subjective quality of life in a rigorous fashion, theorizing about concrete conditions that determine such differences, and testing the resulting empirical predictions (for reviews, see Clarke et al. 2010; Diener and Suh 2000; Frey and Stutzer 2002; Layard 2005).[1] We do so by examining how life satisfaction across the industrial democracies corresponds with different outcomes in the conflict of "politics versus markets". We thus hope to understand how state intervention to "protect" citizens against pure market forces affects the overall quality of human life, using the extent to which people enjoy their lives as the appropriate evaluative metric. To anticipate our findings,

[1] The intellectual infrastructure for studying subjective well-being is sufficiently developed and familiar so as not to require extensive elaboration. A voluminous literature has documented that conventional survey items utilized to measure subjective well-being are reliable and valid (for a discussion, see Myers and Diener 1995). After an exhaustive review, Veenhoven concludes that any misgivings about measurement "can be discarded" (Veenhoven 1996, p. 4). Similarly, the collective evidence strongly endorses the proposition that linguistic or cultural barriers (including social pressures for over—or under-reporting self reported satisfaction) do not meaningfully detract from our ability to make cross-national comparison (see, for example, Inglehart 1990 and Veenhoven 1996, 1997a, b). More recent literature equally supports this position. Kacapyr (2008), for instance, argues that subjective well-being indicators do warrant the confidence that the research has placed on them. Another literature, again conveniently summarized by Veenhoven (2002), convincingly argues for the theoretical appropriateness of subjective measures of quality of life, such as satisfaction, as opposed to purely objective indicators (such as income or other measures of consumption).

we find that life satisfaction varies directly with the extent of such protections, net of economic, social, and cultural factors.

The chapter is organized as follows. We first articulate the basic theoretical debate between the market and the democratic process as agents of human well-being, before turning to an appraisal of the existing evidence on how reliance upon these two mechanisms differentially affects human well-being. We then articulate our research design and discuss the empirical results. We close with a discussion of the implications of the findings for our appraisal of the markets and the study of life satisfaction.

12.2 Markets Against Politics

It is widely agreed that the most basic and persistent axis of political and ideological conflict in the industrial democracies is that of the nature and extent of public intervention into the market. Within the political economy literature, this conflict is typically described as one of markets versus politics (e.g., Lindblom 1977). As these are also the two fundamental mechanisms through which well-being can be both produced and distributed (Esping-Andersen 1990), they are the natural locus of attention for those seeking to understand how different political outcomes may affect quality of life.

At the most basic level, the issue at hand is whether to leave the generation and allocation of well-being to the "invisible hand" of the capitalist economy, or to make it at least in part subject to the political decisions of voters. Those favoring the latter ultimately do so because, as Lane puts it, markets are "indifferent to the fate of individuals" (Lane 1978, p. 13). Esping-Andersen summarizes the argument perfectly when he notes that while capitalism certainly has many positive aspects that doubtless do contribute to quality of life, in the end "the market becomes to the worker a prison within which it is imperative to behave as a commodity in order to survive" (Esping-Andersen 1990, p. 36). As it is not controversial to suggest that human beings do not enjoy being reduced to a commodity, it seems equally unremarkable to suggest—*if we accept the metaphor*—that people's lives are likely to be less rewarding the more they are subject to the insecurities inherent in the market. Put differently, the more individuals are "decommodified" by social policy, the greater should be their well-being, to the extent that the critics of markets are correct in their socio-analysis of capitalism.

The counter-arguments are equally straightforward. Two are especially worthy of note (for a review of others, see Veenhoven 2000). The first, most familiar to students of political economy, is the conventional one of the "unintended consequences" of the welfare state specifically, and by extension, other interventions in the market designed to protect workers. Such claims are animated by sophisticated (if not universally accepted) economic theory and (equally disputed) empirical evidence. If the defenders of unfettered markets are right, we should observe a negative relationship between market interventions and human happiness, in that "decommodification" is argued to be merely an ideological mask for deviations from market "efficiency," which will impose itself as costs on the population by inhibiting economic growth, fostering unemployment, lowering wages, and so on, so as to lower the general level of happiness.

Beyond the immediate effects of economic inefficiency, the state's efforts at redistribution and provision may fail, it is argued, because they actually reduce both the "quantity" and "quality" of well-being relative to markets because of their secondary effects on society. These are principally the displacement of the church and family as sources of the traditional source of income support, such that the welfare state and labor market regulations that are intended to decommodify citizens have the effect of severing the communitarian and emotional support that these traditional sources also provided. Similarly, it is often argued that reliance on the state encourages "collectivization" with deleterious consequences for individual privacy, freedom, and autonomy.

Another related but logically distinct line of argument popularized by Murray (1984) and analyzed as a more generalized "ideational" phenomenon dating from the nineteenth century by Hirschman (1991), is what has become known as the "perversity thesis": efforts to ameliorate problems created by the market in turn create "perverse incentives" of a purely moral nature. In this approach, market interventions designed to insulate individuals from the market are construed as imposing moral costs on society, sometimes expressed in ways that do not lend themselves to ready falsification using conventional economic indicators. Somers and Block (2005), for instance, document the way in which this approach, utilized in both England during the debate over the 1834 "New Poor Law" and in the United States during the Reagan years, was deployed by elites to suggest that income maintenance programs induced "laziness" and "degradation" among clients of the welfare state (and for society more generally, since, abstractly, welfare always exists as an option once introduced). As a consequence, again, the greater the level of political intrusion into the market system, the less satisfying life becomes, in this interpretation.

These abstract arguments reduce to an obvious, tangible question: do interventions in the market designed to protect citizens against the insecurity and inequality of the market ultimately contribute to greater or lesser levels of subjective well-being? The most obvious and important kind of such intervention is the welfare state, in the narrowest sense of the social "safety net." Still, two additional issues warrant attention if one is going to consider the general effect of pro- and anti-market public policies. One is the size of the state sector aside from the transfer payments that are the purview of the welfare state, i.e., the share of the economy "consumed"—i.e., controlled by—the state. If the state is conceived of as the locus of democracy, then the amount of government consumption reflects the degree to which the economy is subject to at least nominal democratic authority. If we wish to understand how much "a program of emancipation from the market" affects well-being, we must consider the extent to which the state has in fact displaced the market by considering its economic "footprint" in this way.

Other issues of enormous significance are additional public policies that provide protection against the "insecurity" inherent in a market system (where, again, workers are argued to be "commodified" and thus to see the quality of their lives varying, similar to the way other commodities vary in value). While the size or generosity of the welfare state does certainly provide some evidence on this score, it is also appropriate to consider the direct impact on satisfaction of a measure of insecurity-in-the-labor-market in a more precise fashion, as we attempt below, by employing labor market regulation as an explanatory variable.

12.3 Data and Method

As is conventional in the emerging literature on the cross-national determinants of life satisfaction, we rely upon the pooled World Values Survey, which provides survey data with representative national samples for all OECD countries.[2] Our dependent variable, *life satisfaction*, is the standard question: "All things considered, how satisfied are you with your life now?" There are ten response categories, with higher values suggesting greater satisfaction. We analyze these data in two ways. First, we simply rely on the satisfaction item noted above, modeling it as a function of both individual- and national-level factors. Second, we regress satisfaction on a set of purely individual-level characteristics, and then use the mean of the residuals from this model, by country, as our dependent variable. The logic of the first approach is self evident; for the latter, the goal is to isolate the variation in quality of life across states that cannot be attributed to individual-level phenomena. In the first approach, the unit of analysis is the individual; in the latter it is the state mean.[3]

For both sets of analyses, our principal independent variables are alternative measures of the central theoretical construct of the degree of state intervention in

---

[2] Countries in our study include: Australia, Austria, Belgium, Canada, Denmark, Finland, France, Germany, Great Britain, Greece, Ireland, Italy, Japan, The Netherlands, New Zealand, Norway, Portugal, Spain, Sweden, Switzerland, and The United States.

[3] DiTella et al. (1997) referred to the residual method, when applied to nation states, as providing an estimate of a country's "pure" level of satisfaction, meaning the amount of satisfaction that can be attributed to national characteristics per se. Radcliff (2001), who uses the same method, notes that this may overstate the case somewhat, since politics can influence at least some variables (e.g., income) that are potentially affected to at least some degree by government policy. Still, the procedure is useful because it is highly conservative, given that it assigns as much variance as possible to individual-level factors that are assumed to be apolitical, thus "raising the bar for showing that politics indeed affects average levels of satisfaction" (Radcliff 2001). This approach is familiar to students of American politics in the work of Erikson et al. (1993), who utilize the same logic to estimate states' ideological cultures, i.e. they regress individual ideological orientations on demographic variables, and then include dummies for states, which in turn become their estimates of culture in the sense of being that variance not explained by individual-level variables. We use mean residuals (Radcliff 2001) instead of the value of the dummy variables, but the procedures are econometrically equivalent.

the market. One is the size of government, defined as the state consumption share of real per capita GDP (Penn World Table 6.2). Higher values indicate, of course, a large state sector, or a more "socialistic" economy. We thus expect, if the argument that politics trumps markets in the production of well-being is correct, a positive relationship between this variable and life satisfaction.

Another is the degree of *labor market regulation*, for which we rely upon a scale developed by the Fraser Institute, which measures the extent to which the economy is affected by "mandated minimum wages," the degree to which "the hiring and firing of workers is impeded by regulations," the extent to which wages are "set by centralized collective bargaining," and the degree of governmentally mandated "requirements for advance notice, severance payments, and penalties due when dismissing a redundant worker". As the index is construed by its authors as a measure of economic "freedom," which is to say the absence of labor market regulation, higher values indicate less regulation (i.e., more "freedom").[4] We thus expect it to have a negative relationship with satisfaction.

The above indicators are, as we have argued above, potentially better measures of the extent to which a country's economy is subject to political regulation and control, relative to the free market ideal, than are the measures of welfare state generosity that has been the focus of prior research. That said, we by no means deny the relevance of the welfare state, and thus include an appraisal of the effect of the welfare state on satisfaction in our analysis. Measurement here becomes marginally more complicated. As is now widely accepted, welfare spending, however widely used, is not the best indicator (e.g., Esping-Andersen 1990). Scholars seem to generally agree that more elaborate indices of the level of the decommodification of labor are superior (for an accessible but detailed summary, see Allan and Scruggs 2004). We rely upon the most comprehensive measure of welfare state generosity available, the time-serial extension of the original Esping-Andersen (1990) *decommodification* index developed by Scruggs (2005).[5] As

---

[4] Details in the coding can be found in Appendix 1: Explanatory Notes and Data Sources, pp. 192–193 in the 2008 Economic Freedom of the World Report (http://www.freetheworld.com/2008/EFW2008App1.pdf).

[5] The details of the rigorous operationalization are not readily summarized. Perhaps the most succinct description is offered by Messner and Rosenfeld (1997, p. 1399): the index "encompasses three primary dimensions of the underlying concept: the ease of access to welfare benefits, their income-replacement values, and the expansiveness of coverage across different statuses and circumstances. A complex scoring system is used to assess (the amount of decommodification provided by) the three most important social welfare programs: pensions, sickness benefits, and unemployment compensation. The scoring system reflects the 'prohibitiveness' of conditions for eligibility (e.g., means testing), the distinctiveness for and duration of entitlements (e.g., maximum duration of benefits), and the degree to which benefits replace normal levels of earnings. The indices for these three types of…programs are then aggregated into a combined (additive) index." It should be noted that the individual indices are weighted by the percent of the relevant population covered by the given programs. Each dimensional index is built from multiple indicators (e.g., five for old age pensions, four each for sickness and unemployment) reflecting the concerns noted above. The data can be located at http://sp.uconn.edu/~scruggs/wp.htm (accessed on April 15 2008).

larger values indicate greater decommodification, the predicted relationship with subjective well-being is positive.

In specifying control variables and estimation technique at the individual-level, we have to be sensitive to the effects such choices make on the size of the sample, principally in the sense of affecting the number of countries which may be included in the analysis. Some individual-level variables are simply not available for all nations, such that as we expand the number of individual-level determinants the more countries are excluded. We thus rely on a model that contains the minimal number of demographic controls, which minimizes the amount of missing data. Similarly, we rely on country dummy variables as national-level control variables. This is the most econometrically powerful method for controlling for the pooled structure, accounting as it does for the relatively fixed, social, economic, and cultural characteristics of a given country. The effect of the dummies is, of course, to fit separate intercepts for each country, thus accounting for the large and sustained differences in satisfaction that one might expect to result from different cultural and economic contexts.

When using nations as the units of analysis, country dummies are obviously inappropriate (there being just one observation per country). In this case, we substitute for the dummies a set of substantive national-level controls. Inglehart (1988, 1990, 2000), Inglehart and Klingemann (2000) and Inglehart and Welzel (2005) have written extensively on the role democracy plays in affecting levels of happiness. Though the direction of causality is disputed, it seems clear that democracy is associated empirically with levels of well-being. The importance of a nation's level of economic development, as well as its short term level of economic prosperity (particularly its level of unemployment) have been well-documented as influencing levels of subjective well-being cross-nationally (e.g., Frey and Stutzer 2002). Finally, it seems clear that the extent of "individualization" (Veenhoven 1999) present in national cultures is a consistent predictor of national levels of satisfaction (see also, Schyns 1998). To account for these factors we include, as a measure of democracy, the cumulative Polity democracy scores from 1970 to the year of the observation,[6] real GDP per capita (from the Penn World Tables 6.2), the unemployment rate (from the World Bank's World Development Indicators),[7] and the index of the *individualism* of culture devised by Triandis (1989), data from Diener et al. (1995).

[6] The Polity data set examines concomitant qualities of democracy and autocracy in governing institutions and spans a range of fully institutionalized autocracies through "mixed" or "incoherent authority" regimes, to fully institutionalized democracies. The Polity score captures this regime authority spectrum on a 21 point scale ranging from −10 (hereditary monarchy) to +10 (consolidated democracy). Full details may be found at the Polity data set website (http://www.systemicpeace.org/polity/polity4.htm).

[7] Unemployment data are from the World Development Indicators 2006 edition CD-ROM, from the World Bank. (For more information, see: http://publications.worldbank.org/ecommerce/catalog/product?item_id=5612167).

Research on individual-level determinants of subjective well-being consistently shows that the same basic characteristics tend to affect individuals similarly across countries. Following the conventions of the literature we treat life satisfaction as a function of *gender*, *age*, *age-squared* to account for the curvilinear relationship between age and life-satisfaction, household *income*, whether the chief wage earner is *unemployed*, whether the respondent is *married* (or living as married), whether the respondent has *children* living at home, and frequency of *attending religious services*. Finally, we include as a control the year of the survey, to account for the possible downward secular trend in satisfaction argued to have occurred over recent decades (see especially, Lane 2000).

For the individual-level models, estimation is with Huber-White robust standard errors, correcting for the pooled structure of the data (i.e., country-clustered). This procedure yields estimates that are robust to both between-country heteroskedasticity and within-country correlation. For the country-level analysis, we rely on conventional OLS.

12.4 Analysis

The base results, using individuals as the units of analysis in the fixed effects model, are provided in Table 12.1.

Table 12.1 Life satisfaction and Government

	(a)	(b)	(c)
Size of Government	0.023[c] (0.007)	n/a	n/a
Labor market regulation	n/a	−0.026[a] (0.014)	n/a
Decommodification	n/a	n/a	0.060[c] (0.009)
Gender	0.022 (0.01562)	0.023 (0.01562)	−0.010 (0.017)
Age	−0.051[c] (0.003)	−0.051[c] (0.003)	−0.052[c] (0.003)
Age-squared	0.001[c] (0.000)	0.001[c] (0.000)	0.001[c] (0.000)
Marital status	0.518[c] (0.019)	0.518[c] (0.019)	0.507[c] (0.020)
Children	−0.010 (0.006)	−0.010[a] (0.006)	−0.003 (0.007)
Unemployed, head of house	−0.847[c] (0.047)	−0.845[c] (0.046)	−0.851[c] (0.049)
Income	0.078[c] (0.003)	0.078[c] (0.003)	0.0758[c] (0.003)
Attends religious services	0.063[c] (0.003)	0.063[c] (0.003)	0.07160[c] (0.003)
Year	0.085[c] (0.009)	0.075[c] (0.007)	0.05338[c] (0.009)
Constant	7.510[c] (0.152)	8.066[c] (0.101)	6.517[c] (0.250)
Adj. R-squared	0.15	0.15	0.16
N	59,014	59,014	50,209

Dependent variable is life-satisfaction (1–10 scale). Estimation is with Huber-White country-clustered robust standard errors. Table omits country dummies. Entries are unstandardized regression coefficients (standard errors)

[a] significant at 0.05 level, [b] significant at 0.01 level, [c] significant at 0.001 level

As is apparent, for the size of government (column a), the degree of labor market regulation (column b), and decommodification (column c), the coefficients are significant and of the expected signs. We thus confirm the essential point that the greater the state interventions against the market in the form of the decommodification of labor, the more satisfied citizens tend to be, other factors being equal. The effects of the size of government, as well as the degree of labor market regulation, are also as predicted by the proponents of state intervention. In regard to the former, the greater the size of the state sector, the greater is the degree of satisfaction. Thus, well-being appears to increase as the portion of the economy that is publicly controlled increases. Similarly, well-being decreases as the degree of economic "freedom" increases (hence the negative coefficient), but this is only to say that well-being increases as the amount of regulation of the labor market increases. Put differently, satisfaction with life declines as the economy is less supported by minimum wages, as workers are less protected from dismissal by regulation, provided less notice or severance pay when dismissed, and when their wages are less determined by collective bargaining. Collectively, the empirical connection between the three variables reported in Table 12.1 and life satisfaction provide strong evidence in support of the idea that market interventions improve quality of life.

Turning to the second estimation approach discussed earlier, we regress satisfaction on the individual-level variables used above, and then create a new dependent variable that is the mean value of the residuals, by country. Where the unit of analysis previously was the individual, it is now the nation-state. This method has the advantage, beyond the substantive point of purging the data of entirely individual-level effects that might be argued to be theoretically unrelated to political conditions, of offering an agreeably simple and straightforward econometric analysis.

Table 12.2 presents the results of OLS estimation, in which the residual level of satisfaction is modeled as a function of the mean level of the national-level variables used previously. For labor market regulation (column b) and decommodification (column c), results are substantively identical to the individual-level analysis, with the coefficients of interest being significant and correctly signed. The initial result (reported in column a) for the size of government is of the expected sign, but is not strictly significant. However, an inspection of the df-betas (a statistic for assessing the potential of single cases to have an exceptional influence on the overall results) suggests only one case that exceeds the critical value (France); excluding that single case returns the coefficient to significance.[8] The aggregate-level analysis thus clearly confirms the earlier findings for the welfare state and regulation, and provides at least qualified support for the size of the state sector.

[8] Note again that this is the only case suggested by the diagnostic. Other conventional regression diagnostics suggest nothing untoward.

Table 12.2 Life satisfaction and Government. Aggregate analysis

	(a)	(b)	(c)
Size of Government	0.015 (0.012)	n/a	n/a
Labor market regulation	n/a	−0.095[a] (0.045)	n/a
Decommodification	n/a	n/a	0.027[a] (0.012)
GDP per capita	0.000 (0.000)	0.000 (0.000)	0.000 (0.000)
Unemployment	−0.032 (0.026)	−0.038 (0.023)	−0.022 (0.036)
Individualism	0.121[a] (0.162)	0.127[a] (0.053)	0.113[a] (0.055)
Democracy	0.003 (0.002)	0.003 (0.002)	0.006[a] (0.003)
Constant	−0.712 (0.927)	−0.090 (0.603)	−2.46[a] (1.33)
Adjusted R-squared	0.33	0.40	0.47
N	19	19	17

Dependent variable is mean life-satisfaction (1–10 scale). Estimation is with OLS. Entries are unstandardized regression coefficients (standard errors)
[a] significant at 0.05 level, [b] significant at 0.01 level, [c] significant at 0.001 level

Perhaps the most profitable way of interpreting the substantive, rather than merely statistical, significance of the theoretically relevant variables is to compute the expected change in residual satisfaction (for convenience, expressed in standard deviations) when moving from the minimum to the maximum observed values of the former variables. Moving across the range of decommodification suggests a predicted change of 1.36 standard deviations in the life satisfaction. Moving across the range of the labor market variable has a similar effect (1.30 standard deviations). The effect of the size of government, using the reported coefficient, is somewhat smaller (0.78 standard deviations), but is comparable to the other results when removing the leverage point noted above (1.35 standard deviations). The real world impact of the market variables is thus quite substantial.

12.5 Discussion

The principal empirical conclusions emerging from the analysis are clear: life satisfaction varies directly with the size of government, the generosity of the welfare state, and the degree of pro-worker state regulation of the labor market. In elaborating the implications of these results it may be helpful to begin by noting what they do *not* imply. First, they do not settle older and broader questions about the future of the welfare state writ large (see Swank 1988, 2001). They do not provide any overall judgment on whether generous welfare policies are good or bad; whether regulating labor markets to protect workers is inherently good for society; or that a larger state sector is objectively superior to a smaller one. These questions are inherently both normative and ideological. As such, they do not have empirical "answers." We make no pretense of offering any.

That said, our results, taken on their face, certainly do have implications for our empirical understanding of subjective well-being. However, we would again begin by noting what they cannot be construed as suggesting, viz. that a market economy is inimical to well-being. On the contrary, it seems certain that capitalist economies are superior producers of well-being than prevailing non-market alternatives (e.g., Veenhoven 2000). Thus, the essential point of the analysis is not that the market inhibits well-being, but rather that, within the context of a capitalistic economy, political interventions that attempt to redress market deficiencies tend to produce greater levels of human happiness. Our results thus do not indict the market as it affects satisfaction with life, but suggest instead that the quality of human life is best when the inequalities and uncertainties of the market are mitigated by state intervention acting in the interests of workers and citizens. Our results might thus be most easily summarized by suggesting that it is "compassionate capitalism" that seems most consistent with well-being.

More generally, our fundamental conclusion is that politics emphatically does matter for what is arguably the most fundamental issue in social science: identifying the conditions that make human life rewarding. Such a contention must be considered good news for proponents of democracy, whatever their ideological preferences. The choices made by voters in choosing governments, and the subsequent policy decisions those governments undertake, do have important consequences. Democracy, then, matters.

Our findings also have implications for the academic study of subjective well-being. Most obviously, we offer further evidence in support of the disputed contention that governmental policies affect quality of life. More importantly, perhaps, this fact in turn has implications for our theoretical understanding of what determines well-being. We would argue that the evidence presented here suggests more than we add another set of variables to a list of those thought to affect quality of life. By demonstrating that public (i.e., democratic) "intrusion" into the market improves life satisfaction, we hope to focus scholarly attention on the basic question of theoretical approaches to modeling the determinants of well-being. The conventional approach in psychology and economics is implicitly, and perhaps unconsciously, to assume that society is composed only of individual persons, who happen to vary in their many individual-level characteristics but who remain largely undifferentiated by macro-level conditions aside from (a) the level of affluence and (b) culture. Thus, in the much cited, nearly encyclopedic review of the "Three Decades of Progress" in the study of subjective well-being by Diener et al. (1999) these are the only two societal factors discussed. To be sure, more recent work, reviewed previously, has touched upon macro-conditions, but the fact remains that far too little attention has been devoted to theorizing about how socio-political conditions determine quality of life. In demonstrating the importance of political outcomes, we highlight the need for richer theories that incorporate such factors.

The present study may also point toward the direction such theorizing might take. By illustrating that welfare spending, labor market regulation, and other political interventions into the economy affect well-being, we also suggest the centrality to human life of the market economy itself. As Lindblom (1977) has persuasively argued, we tend as social theorists to take the market for granted, in the sense of considering it to be a fixed characteristic—almost a natural force of nature, akin to gravity. Instead, we need to be cognizant of the fact that the market is a variable, in the sense that it varies both in its existence but also in its character. There are, as is commonly accepted, different "flavors" of capitalist democracy (e.g., Esping-Andersen 1990; Huber et al. 1992). Variations in the nature of the market system across time and space would appear to be essential elements in any understanding of life satisfaction.

Similarly, it seems highly likely that the market system itself, as the central institution of contemporary society, warrants greater attention in the study of subjective well-being (Radcliff 2013). It takes no great insight to suggest that any theory of human well-being should include the nature and logic of the market as explanatory factors. We thus conclude with the contention that our understanding of quality of life and the mechanisms that enhance or diminish it must be based upon an understanding of how the market system, as the principal institution structuring the modern world, affects the happiness of the people whose lives it largely defines.

References

Allan J, Scruggs L (2004) Political partisanship and welfare state reform in advanced industrial societies. Am J Polit Sci 48(3):496–512

Clarke HD, Whiteley P, Sanders D, Stewart M (2010) Government performance and life satisfaction in contemporary Britain. J Politics 72(3):733–746

Diener E, Diener M, Diener C (1995) Factors predicting the subjective well-being of nations. J Pers Soc Psychol 69:851–864

Diener E, Suh EM, Lucas RE, Smith HE (1999) Subjective well-being: three decades of progress. Psychol Bull 125:276–302

Diener E, Suh EM (eds) (2000) Culture and subjective well-being. MIT Press, Cambridge

DiTella R, McCulloch R, Oswald A (1997) The macroeconomics of happiness. The labour market consequences of technical and structural change discussion paper series, number 19. Centre for Economic Performance, Oxford University

Erikson RS, Wright GC, McIver JP (1993) Statehouse democracy: public opinion and policy in the American states. Cambridge University Press, Cambridge

Esping-Andersen G (1990) The three worlds of welfare capitalism. Princeton University Press, Princeton

Frey B, Stutzer A (2002) Happiness and economics. Princeton University Press, Princeton

Hirschman A (1991) The rhetoric of reaction: perversity, futility, jeopardy. Harvard University Press, Cambridge

Huber E, Ruesschemeyer D, Stephens J (1992) Capitalist development and democracy. University of Chicago Press, Chicago

Inglehart R (1988) The renaissance of political culture. Am Polit Sci Rev 82(4):1203–1230

Inglehart R (1990) Culture shift in advanced industrial democracies. Princeton University Press, Princeton

Inglehart R (2000) Globalization and post-modern values. Wash Q 23(1):215–228

Inglehart R, Klingemann HD (2000) Genes, culture, democracy, and happiness. In: Diener E, Suh EM (eds) Culture and subjective well-being. MIT Press, Cambridge, pp 165–183
Inglehart R, Welzel C (2005) Modernization, cultural change, and democracy: the human development sequence. Cambridge University Press, New York
Kacapyr E (2008) Cross-country determinants with life. Int J Soc Econ 35(6):400–416
Lane RE (1978) Autonomy, felicity, futility. J Politics 40:1–24
Lane RE (2000) The loss of happiness in market democracies. Yale University Press, New Haven
Layard R (2005) Happiness: lessons from a new science. Allen Lane, London
Lindblom C (1977) Politics and markets. Basic Books, New York
Messner SF, Rosenfeld R (1997) Political restraint of the market and levels of criminal homicide: a cross-national application of institutional-anomie theory. Soc Forces 75(4):1393–1416
Murray C (1984) Losing ground: American social policy, 1950–1980. Basic Books, New York
Myers D, Diener E (1995) Who is happy? Psychol Sci 6:10–19
Radcliff B (2001) Politics, markets, and life satisfaction: the political economy of human happiness. Am Polit Sci Rev 95(4):939–952
Radcliff B (2013) The political economy of human happiness: how voters' choices determine the quality of life. Cambridge University Press, Cambridge
Scruggs L (2005) Comparative welfare entitlements dataset. Department of Political Science, University of Connecticut. Available at: http://sp.uconn.edu/~scruggs/wp.htm. Accessed 15 April 2008
Schyns P (1998) Cross-national differences in happiness: economic and cultural factors explored. Soc Indic Res 43:3–26
Somers M, Block F (2005) From poverty to perversity: ideas, markets, and institutions over 200 years of welfare debate. Am Sociol Rev 70(2):260–287
Soros G (1998) The crisis of global capitalism. Public Affairs, New York
Swank D (1988) The political economy of government domestic expenditure in the affluent democracies, 1960–1980. Am J Polit Sci 32(3):1120–1150
Swank D (2001) Political institutions and welfare state restructuring: the impact of institutions on social policy change in developed democracies. In: Pierson P (ed) The new politics of the welfare state. Oxford University Press, New York, pp 197–237
Triandis HC (1989) The self and social behavior in differing cultural contexts. Psychol Rev 96(3):506–520
Veenhoven R (1996) Developments in satisfaction research. Soc Indic Res 37:1–46
Veenhoven R (1997a) Advances in understanding happiness. Revue Quebecoise de Psychologie 18:29–74
Veenhoven R (1997b) Quality of life in individualistic societies. In: DeJong M-J, Zijderveld AC (eds) The gift of society. Enzo Press, Netherlands, pp 149–170
Veenhoven R (1999) Quality of life in individualistic society: a comparison of 43 nations in the early 1990s. Soc Indic Res 48:157–186
Veenhoven R (2000) Well-being in the welfare state: level not higher, distribution not more equitable. J Comp Policy Anal 2(1):91–125
Veenhoven R (2002) Why social policy needs subjective indicators. Soc Indic Res 58:33–45

Chapter 13
Should the State Care for the Happiness of its Citizens?

Aloys Prinz

13.1 Introduction

Should the state care for the happiness of its citizens? Taking a look at the most recent developments in a number of countries (as, for instance, from Australia to France, from South Korea to the UK) and in international organizations (see, e.g., the OECD website on "The Measuring of Well-being and Progress" at www.oecd.org/measuringprogress), the answer seems to be: yes, of course, or as the OECD expressed it: "Improving the quality of our lives should be the ultimate target of public policies" (Angel Gurria, Secretary-General of the OECD; see OECD 2011). Although the concept of "quality of life" is not the same as the concept of "happiness", this quotation shows a tendency to employ subjective indicators as public policy goals (as supplements or as alternatives to "objectively measurable" entities like, e.g., GDP).

But not all persons are convinced that the state should be responsible for the happiness of its citizens. One of the most prominent opponents was presumably the philosopher Karl Popper who wrote: "But of all political ideals, that of making the people happy is perhaps the most dangerous one" (Popper 1974, p. 237). This is in stark contrast to the now seemingly dominant public opinion. These differing positions on the relevance of happiness for politics and public policy require an in-depth examination. The most important questions are: How can the state know what makes its citizens happy? And if it knows that, will the state be able, willing and well-equipped to make its citizens happy or at least happier? As we will recognize very soon, these questions are neither new nor are they easy to answer. Still, happynomics[1]—a merger of economics and psychology—provides some tentative clues.

[1] See Conway (2009) for this neologism.

A. Prinz (✉)
Institute of Public Economics, Münster School of Business and Economics,
University of Münster, Münster, Germany
e-mail: A.Prinz@wiwi.uni-muenster.de

H. Brockmann and J. Delhey (eds.), *Human Happiness and the Pursuit of Maximization*, Happiness Studies Book Series, DOI: 10.1007/978-94-007-6609-9_13,

The paper is structured as follows: In Section Two it will be shown that the discussion on the proper role of the state in respect of the well-being of its citizens goes back in Europe at least to the ancient Greeks. It is argued that the discussion about the role of the state has never ended, with different answers being given at different points in time. The arguments in favor of and against happiness as an objective of public policy during the recent debate will be discussed in the third section. An analysis of what governments can do to make people happier is presented in Section Four. The fifth section concludes.[2]

13.2 The Ancient Roots of Today's Discussion

As is well-known, the roots of the culture of the western world can be traced back to ancient Greece (McMahon 2004, 2006). Unsurprisingly, this is also the case with respect to philosophizing about happiness and the ideal state. The philosopher Plato (about 380 BC) described in his "Politeia" ("The Republic") such an ideal state in which everything is organized and enforced by the state. This state is a communist institution completely lacking individual freedom. Although this vision of a perfect state is rather terrifying according to contemporary western standards of civil freedom, the philosopher-king ruling this state should be recognized as a benevolent dictator. However, even if one were to accept that one single person would be able to know what makes people happy, the methods described in the "Politeia" to enforce the vision of the ideal state would nowadays lead presumably to extreme unhappiness among its citizens. Karl Popper warned exactly against such a state, admittedly against the historical background of two totalitarian regimes, the Nazi regime and post-war communism.

The second Greek philosopher who is highly important in contemporary research on happiness is Plato's disciple Aristotle. He is the author who introduced in his work "Nicomachean Ethics" one of the most fundamental differentiation of notions in happiness research by separating *hedonic* happiness from *eudaimonic* happiness [Aristotle 350 B. C. (1962)]. Hedonic happiness means the happy moments in life which we are looking for, whereas eudaimonic happiness means the good life (contemporarily typically interpreted as "satisfaction with life as a whole").[3] Further, Aristotle seems to understand happiness as an individual and personal objective; nevertheless, he recognized the Greek form of democracy, the "Polis", as best suited for providing the conditions for a happy (and virtuous) life. The lesson to be learnt from Aristotle is that, first of all, life is more than having fun. The ethical aspect of life—and also one's own judgment of life—is crucial for

---

[2] In this chapter, my own results of earlier research on happiness are used; see especially Prinz (2009), Bünger and Prinz (2010), as well as Prinz and Bünger (2011a, b, 2012).

[3] Obviously, hedonic happiness can hardly be influenced by public policies or institutions. This kind of happiness is not considered in this chapter.

the good life, i.e., satisfaction with life as a whole (see Sparshott 1994, for a comprehensive comment on Aristotle's "Nicomachean Ethics"). The role of the state is to provide the requirements of a good life, whereas the individual is responsible for the realisation of a good and happy life, and not the state.

The third ancient philosophical perspective on happiness is the approach of Aristippus of Cyrene and Epicurus. Altough being often labeled as proponents of hedonic happiness only, they suggested ethical hedonism: a life full of joy, without pain and without fear of death and without fear of the gods (Epicurus, 341–270 BC). Happiness was considered to be an individual's personal phenomenon, without an explicit role for the state. The lesson to be learnt from these philosophers is that the most serious aspect of life is to avoid unhappiness in the form of fear and pain.

13.3 Should the State Care? Main Arguments

As early as 1980, Richard Layard (1980), a British economist from the London School of Economics, asked for the use of human satisfaction as guidance for public policy, and his 2005 book on that topic became very popular (Layard 2005). Meanwhile, there exists a large and growing body of academic literature as to whether and, if yes how, the state may use results from happiness research for public policy, and this controversial debate still continues.[4] In the following, the arguments for and against the use of happiness as a guide for public policy are summarized (see also De Prycker 2010).

Pro Arguments

It can indeed hardly be denied that happiness plays a crucial role in any person's life. Hence, happiness should be the ultimate objective of public policy—at least from the view of the proponents of this idea. As argued by Layard (2005), Diener (2006) and others, the measurement of happiness via questionnaires is considered feasible, valid and reliable. Given measurability, the outcome of public policies could be determined via their impact on people's happiness; as a consequence, happiness may be employed as an outcome indicator for public policy measures (Diener 2006; van Praag and Ferrer-i-Carbonell 2004). Moreover, happiness indicators enable the quantification of the so-called intangible effects of policies,

[4] Note that an extensive overview of relevant papers is not intended here. In addition to the papers quoted in the main text, Kahneman et al. (2004), Veenhoven (2004) as well as Ng and Ho (2006), Ng (2008) and Ott (2010) (among others) seem to support the employment of satisfaction with life as a subjective well-being indicator for public policy, whereas Duncan (2010) and Brülde (2010) (among others) seem to be critical in this respect.

e.g., in areas of health care, social services and the environment (Diener 2006; Diener et al. 2009; van Praag and Ferrer-i-Carbonell 2004).

As revealed by behavioral studies in economics, there seem to be certain decision problems where individuals fail because of myopia or weakness of will (Laibson 1997) and strong information deficits. These are especially decisions with a rather long time horizon, such as saving for old age, and decisions concerning so-called credence goods (see Dulleck and Kerschbamer 2006, and the literature quoted therein). The latter are goods whose quality cannot be determined upon inspection or even experience, but which require expert knowledge which lay-persons usually do not have. Nowadays the number of these goods has increased, among them pharmaceuticals and the like which are heavily regulated, but also supposedly simple things, such as food. Due to high-tech innovations hardly any consumer knows what exactly he or she is eating. Markets do not seem sufficiently well-equipped to direct demand towards what is best for consumers (for an economic analysis see, e.g., Dulleck and Kerschbamer 2006, Wolinsky 1995). As a consequence, malnutrition occurs more often than in the past; this has led to an increase in the number of people suffering from, e.g., obesity and diabetes. Public policy could be guided here to take adequate measures to fight market failure with credence goods. The question of *whether* public policy should concern itself with credence goods can be answered without consulting happiness research; by way of contrast, the question on how public policy should intervene may employ results from happiness research (in addition to or in combination with results from behavioral economics). Most people dislike being patronized by the state (e.g., Sunstein and Thaler 2003, 2006). Autonomy and freedom of choice seem to be important for people's well-being; on the other hand, the same holds true for security and predictability. As a consequence, successful public policies should support people in prudent decision-making without patronizing them (Sunstein and Thaler 2003, 2006; Thaler and Sunstein 2003, 2008).

Contra Arguments

The first argument against happiness as the ultimate objective of public policy is the questionable measurability of happiness (in the sense of subjective well-being). Supporters of a happiness-oriented public policy assert that this question is ultimately answered. However, reviewers of the measurement instruments argue that questionnaires are prone to several shortcomings; this is because, e.g., the responses depend on the interview context (Schwarz 1999, 2003, 2007). All in all, retrospective questions on subjective well-being do not seem to be reliable to the extent that might be necessary for public policy.

Moreover, from a welfare economics viewpoint the status of subjective well-being is unclear; until now, there has been no alternative theory of welfare to neoclassical welfare economics which is commonly accepted (see Gul and Pesendorfer 2007, for a discussion). Probably the most serious criticism of happiness indicators as a guideline of public policy comes from the public choice theory. Social policies are rarely evaluated with respect to their outcomes, although quantitative outcome measures are often available. It seems that politicians are not very much interested in evaluating policy instruments. Although this is not a specific argument against happiness as a policy goal, the argument is nevertheless also valid in this context. The reason is that the objective of politics is rent seeking (McChesney 1997), which is apparently quite different from whatever a welfare economist or a happiness researcher would suggest. Moreover, even if reliable and valid measures of happiness were available and if they were introduced to evaluate the outcome of social and public policies, it is to be expected that politicians would manipulate these measures (Frey and Stutzer 2000, 2010). As is well-known, for instance, the relationship between monetary aggregates and the rate of inflation vanished when they became benchmarks for monetary policy.

To sum up, although there are valid arguments that promote the use of happiness indicators for public policy, the number and seriousness of contra arguments is hardly less convincing.

13.4 Towards a Happiness-Enhancing Welfare State?

Ideal Forms of Welfare State Policies

To start with, a classification of welfare state policies according to their general attitude with respect to *autonomy and choice* on the one hand and to *security and predictability* on the other hand is presented. Both aspects seem relevant here since both are crucial for individual happiness (see Verme 2007, for autonomy and choice, and Pacek and Radcliff 2008a, b, for security and predictability). From an individual perspective, autonomy and choice as well as security and predictability (of the material basis of existence) are required to be able to have some control over one's own life and also to feel comfortable with life. However, although both aspects may go hand in hand in some cases, there are other situations in which more security and predictability mean less or even no autonomy and choice, and vice versa. In the following, a state that considers security and predictability as preferable to autonomy and choice is dubbed a "nanny state"[5]; a state that considers autonomy and choice more important than security and predictability is called a "night-watchman state"[6], and finally a state that considers both aspects more or less equally is denoted as being an "avuncular state".[7]

[5] This notion means that the state as a kind of "nanny" pampers its needy citizens in such a way that they do not have much freedom of choice. The "nanny state" might also be called "paternal socialism" (Buchanan 2005, p. 23).

[6] This notion was created from the German politician Ferdinand Lasalle ("Nachtwächterstaat") in the 19th century to describe a kind of minimal state; see Duden (2009).

[7] The notion of the "avuncular state" was coined by The Economist (2006). See also the "Soft Paternalism" blog of Jones et al. (2009).

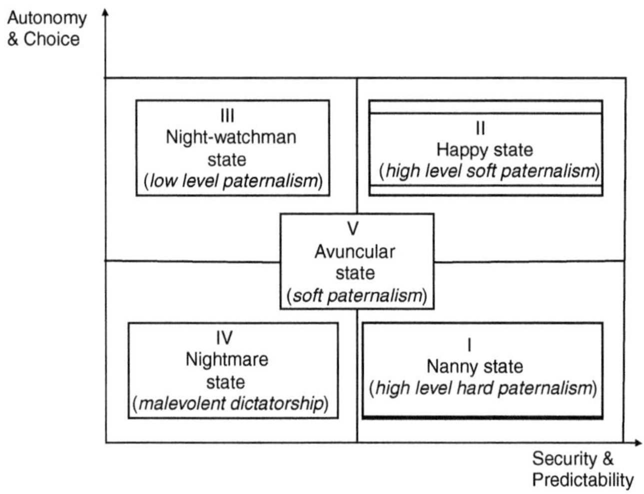

Fig. 13.1 Welfare state regimes. *Source* Own depiction

As depicted in Fig. 13.1, all five (actual and potential) welfare state regimes encompass the *level* of (social) security and protection a state guarantees for its citizens as well as the *tools* it employs to this end, i.e., whether (and to what extent) it uses prescriptions and taxes, etc. (also called "hard paternalism") or "nudges" (Thaler and Sunstein 2008) in the form of default values and the supply of subsidies (also called "soft paternalism"). "Hard paternalism" means a low degree of autonomy and choice for people applying for support from the state, whereas "soft paternalism" means support that leaves a substantial degree of autonomy and choice. However, the price for autonomy and choice is a smaller level of security and predictability and vice versa. Hence, in its ideal form as understood here, the "nanny state" (I) is more generous concerning the level of security and protection and it employs mainly tools of hard paternalism, whereas the "night-watchman state" (III) is less generous, (i.e., it creates more work incentives and forces less redistribution), but it also mainly employs instruments of hard paternalism. The opposites in the other direction (from the southwest to the northeast in Fig. 13.1) are the "nightmare state" and the "happy state" (II). The "nightmare state" (IV) provides neither security and predictability nor autonomy and choice; it can be considered as a malevolent dictatorship. In contrast, the "happy state" combines a high level of security and predictability with a high level of autonomy and choice by using tools from soft paternalism ("nudges") on a high level.

The position in the middle might be awarded to the "avuncular state" (V) with mean levels of security and predictability, combined with a mean level of autonomy and choice which employs the tools of soft paternalism.

The most extreme cases are the "nightmare state" and the "happy state". Whereas a "nightmare state" seems not to be compatible with a free society and democracy, this might be different with the "happy state". A "happy state" not only requires a generous level of income without the obligation to work to ensure a high level of security and predictability, but also a low level of state control to guarantee autonomy and choice. However, as argued by van Suntum (1991), it is not possible to have at the same time generous transfer payments, a low level of state control of individual behavior and resources and a low level of social security fraud. As a consequence, neither the "happy state" nor the "nightmare state" are taken into further consideration here.

Inequality, Social Policy and Happiness

According to the study of Di Tella and MacCulloch (2005), the highest positive effects of macro-determinants of life satisfaction (eudaimonic happiness) in European countries come from GDP per head, the personal income position, and unemployment benefits. The strongest negative effects are due to the unemployment rate, the number of working hours, the inflation rate, being unemployed, living separated from one's partner, and being divorced. Material living conditions as well as the quality of intimate relationships seem to be very crucial to the personal level of satisfaction with life.

One of the most important negative micro-determinants of happiness is unemployment. This confirms the result from the macro-perspective. Although unemployment benefits can compensate to a certain extent for job losses, they are certainly not sufficient to compensate for the reduction of subjective well-being which results from being out of a job (Böckerman and Ilmakunnas 2006; Clark and Oswald 1994, 2002; Di Tella et al. 2001; Winkelmann and Winkelmann 1995, 1998). Therefore, policies directed towards the unemployed apparently increase subjective well-being, perhaps for almost all citizens.

The effects of social security expenditures (relative to GDP) on happiness are not clear in general. Ouweneel (2002), when comparing happiness, life satisfaction and the health of the unemployed in 42 nations in 1990 on the macro-level, could not find empirical evidence for a relation between social security expenditures and subjective well-being. Veenhoven (2000) also reached a similar conclusion when comparing macro-level health and happiness in 41 nations between 1980 and 1990. Furthermore, Veenhoven neither found a cross-national connection between the size of the welfare state (measured by social security expenditures) and average happiness nor between welfare state size and the distribution of happiness. In a longitudinal analysis, neither increases nor decreases of welfare spending did change average subjective well-being. Single-country studies on Sweden (Fors 2010) and Denmark (Greve 2010) came to a similar conclusion.

On the other hand, Pacek and Radcliff (2008b) found empirical support at micro-level for the idea that the degree of "decommodification" affects subjective well-being[8]: the better people can live independent of markets due to social policy the happier and more satisfied people seem to be on average.

A related question is whether (income) inequality is a source of unhappiness in societies. In this respect, there is no clear answer either. Nevertheless, some empirical results are noteworthy. Firstly, Alesina et al. (2004) found some statistical evidence that, especially in Europe, people report a lower level of happiness in combination with greater inequality. Moreover, left-wing persons and poor people in Europe are more negatively affected when inequality is high than persons who are rich or who tend more to the right of the political spectrum (Alesina et al. 2004). A broad international study by Berg and Veenhoven (2010) concluded that there is no negative correlation between income inequality and happiness when controlling for wealth.

Taking the empirical results of happiness research seriously, the question is what kind of welfare state might be best suited when the objective is to make people happier. It should be obvious from the presentation of the nanny state, the night-watchman state and the avuncular state above that neither of these ideal types appears to be well-suited. The nanny state provides a generous level of security and predictability, but at the price of a low degree of autonomy and choice. The reverse is true for the night-watchman state: emphasising autonomy and choice, the level of security and predictability is apparently too low. But also the alternative regime in-between, namely, the avuncular state, has its shortcomings. Since the level of security and predictability is lower than in the nanny state, this might be to the disadvantage of some groups in society whereas for other groups the level of autonomy and choice is still too low.

A Pragmatic Welfare State Regime

The empirical results on the relationship between the welfare state and happiness presented above can be interpreted as follows: On the one hand, guaranteeing citizens a reasonably convenient life independent of the insecurity of markets with respect to unemployment, health care and pensions, makes people more satisfied with their lives. On the other hand, social security expenditures have no happiness-enhancing effect. The conclusion, then, may be that security and protection are happiness-increasing. Nevertheless, financing social policy reduces happiness in the society. Therefore, social security expenditures as such are not happiness-increasing.

Nevertheless, there might be another more pragmatic version of the welfare state that extends the avuncular state, as will be shown in the following.

---

[8] The composite index of decommodification is not easy to interpret. Generally it could be said that a higher value of the index means that persons can live better without being dependent on markets (Pacek and Radcliff 2008b). However, one of the problems of the index is that it does not contain any costs of decommodification.

The most crucial result of happiness research concerning highly developed and rich countries in this respect is that *unemployment* is the single most important determinant of *un*happiness from the macro-level as well as from an individual perspective. As a consequence, the avoidance of unemployment should be the dominant goal of social policy for persons who are fit for work. A welfare state strategy which employs this goal as its dominant objective could be shaped as follows (see, for this strategy and the following, Flaschel et al. 2010, Forum Flexicurity 2008; Funk 2008; Viebrock and Clasen 2009). Firstly, all labor market institutions (labor market laws, social security laws, state agencies, employers, employers associations, and employee associations) should coordinate their labor market-related activities. The next step is to liberalize labor markets by giving up extensive laws that forbid dismissals and that make employment very costly. Although this seems to contradict the aim of securing a very high level of employment, results from economic research appear to suggest that that is exactly what is required.

Since this policy combines flexibility of the labor market with security for the employees, it is called "flexicurity" (see, e.g., Forum Flexicurity 2008). Probably not by accident, Denmark, where this kind of policy is used (Andersen and Svarer 2007), has one of the highest scores for life satisfaction.

A related question is whether research on happiness can tell us something about other areas of social policy such as health care, pensions and poverty. All of these policy areas are relevant for subjective well-being because they are crucial for security and predictability as well as for autonomy and choice. From behavioral economics it is well-known that people are not rational enough when deciding on long-term financial planning and that there might be either too few or too many choices. As pointed out by Laibson (1997), discounting over longer terms leads to inconsistencies; this is because the (implicit) discount rates change over time in such a way that so-called preference-reversals may occur (i.e., savings for old age are used for spending on something else before retirement). The absence of choices, e.g., by forcing people to save for retirement with a public mandatory insurance scheme, may trigger reactance (Brehm and Brehm 1981), i.e., resistance with respect to paying contributions or taxes. Too much choice (Schwartz et al. 2002; Schwartz 2004), as, e.g., with old-age insurance schemes from private firms, might be too complex to understand; consequently, people may abstain from insurance or they may choose a contract that is not adequate.

To overcome or to mitigate these decision-making deficiencies, behavioral economics recommends the introduction of a certain level of freedom of choice, but also the prudent definition of default values which becomes enacted if people do not choose a different option (this soft kind of paternalism is called "nudging"; see Thaler and Sunstein 2008 for a comprehensive description with examples). For instance, if a couple want to marry in Germany, they can either choose the "public" marriage contract as provided by the civil law, or they can negotiate a marriage contract which may contain rules quite different by using the civil law. In regard to the default contract, as provided by law, the politicians should draft it very carefully to ensure that in case of a divorce the couple are not trapped in endless haggling and hassle. The economic reason for the importance of prudent default rules are the information and negotiation costs that otherwise have to be incurred. Moreover, most people shy away from the psychic costs of difficult decisions as, e.g., in the case of a marriage contract (Cigno 2011, p. 30).

The risk is, however, to trust in defaults when they are not adequate to the individual situation. The question remains whether it is possible to define defaults in such a way that nobody is made worse-off. As these doubts remain, defaults should be applied in very important cases only. Besides marriage, the respective areas could include provision for old age, for health care and for long-term care as well as for education. Presumably, and not by accident, these are indeed the areas in which most social policy provisions are currently implemented, but not necessarily with instruments of soft paternalism.

Some new developments concerning tax policy could also be better suited in respect to happiness. To deter tax evasion, almost all countries employ a combination of repressive measures consisting of audit probabilities and punishments. Recent research in tax psychology and economics shows that repressive policies are restricted in their effectiveness because they may undermine people's trust in the government and the tax authorities (Kirchler 2007, p. 202 *et seq.*). When this happens, tax evasion could even increase because of measures which are too repressive. Although some deterrent policy measures are required, governments and tax authorities should adopt a "client and service" position (Alm et al. 2010) to create a trustful relationship with citizens instead of the usual "cops-and-robbers" attitude which produces distrust (Kirchler 2007, p. 189 *et seq.*); trustful citizens will in turn pay taxes voluntarily and happily. As many countries are running out of financial resources, it becomes more important to convince citizens to pay taxes honestly than simply to let them fear the taxman. Tax resistance nowadays means much higher costs of tax collection and an eroding tax base. As in other policy areas too, politicians must convince people to comply with laws they do not agree with. A good policy method is, e.g., to *involve* people as much as possible in designing policies; it is the participation itself that supports subjective well-being (this is called "procedural utility" by Frey et al. 2004; see also Frey and Stutzer 2005; Benz 2007). In general, to get people involved in public policy decisions that are relevant for them seems to be an important approach to make people happier, but it will be difficult to implement. However, as demonstrated in Switzerland, it is not impossible.

A special challenge for social policy as well as happiness policy is poverty. Hardship is generally not compatible with happiness. Providing the socio-economic means for a life without hardship and misery is necessary even from a humanitarian perspective irrespective of other criteria. This is apparently a policy area where all relevant welfare state regimes seem to agree that it is necessary to pay social assistance to those living in poverty and to provide unemployment benefits to the unemployed for a certain time. As pointed out above, a "flexicurity" strategy could even improve the well-being of poor and unemployed persons. The reason is that unemployment is one of main reasons for poverty in rich societies.

13.5 Conclusion

Should the state care for the happiness of its citizens? In the literal sense of the paper's title the question is apparently to negate: No, it does not seem to be the business of the state to make us all happy. The reasons—which are more or less the same as those presented by Popper more than half a century ago—are: (1) despite the results of empirical research on happiness, only the individual can find her or his route to happiness; (2) even if the state knew what makes us individually happy, enforcing this happiness would make us unhappy since autonomy as well as individual freedom and control are crucial ingredients of happiness; and (3) the experiences in the 20th century with totalitarian regimes should make us cautious in respect to the happiness promises made by the state. To this extent, Plato's position is apparently not to be recommended.

As pointed out by Aristotle, eudaimonic happiness requires a stable institutional framework that empowers and supports people to make the best of their lives. It is indeed the business of the state to provide such an institutional framework. The crucial question is whether a happiness-oriented policy would be better suited to design this framework than other orientations from politics, sociology, economics, etc. The negative result of this paper is that there is no pure ideal welfare state regime that seems justifiable by happiness research. However, applying the *principles* of soft paternalism within an avuncular state, supplemented by measures to avoid hardship and unemployment, might be well suited to design a better institutional framework than the existent one. If successfully implemented, this new policy framework would please not only Karl Popper but also Richard Layard.

Acknowledgments Extensive and useful comments to earlier versions of this paper from Jan Delhey and Björn Bünger are gratefully acknowledged. The remaining errors are mine.

References

Alesina A, Di Tella R, MacCulloch R (2004) Inequality and happiness: are Europeans and Americans different? J Public Econ 88:2009–2042

Alm J, Cherry T, Jones M, McKee M (2010) Taxpayer information assistance services and tax reporting behavior. J Econ Psychol 31:577–586

Andersen TM, Svarer M (2007) Flexicurity—labor market performance in Denmark. CESifo Working Paper 2108

Aristotle (350 B. C. [1962]) The Nicomachean ethics—with an English translation by H. Rackham, William Heinemann, London

Benz M (2007) The relevance of procedural utility for economics. In: Frey BS, Stutzer A (eds) Economics and psychology. A promising new cross-disciplinary field. MIT Press, Cambridge, pp 197–228

Berg M, Veenhoven R (2010) Income and happiness in 119 nations: in search for an optimum that does not appear to exist. In: Greve B (ed) Happiness and social policy in Europe. Edgar Elgar, UK, USA, pp 174–194

Böckerman P, Ilmakunnas P (2006) Elusive effects of unemployment on happiness. Soc Indic Res 79(1):159–169
Brehm SS, Brehm JW (1981) Psychological reactance: a theory of freedom and control. Academic Press, New York
Brülde B (2010) Happiness, morality, and politics. J Happiness Stud 11(5):567–583
Buchanan JM (2005) Afraid to be free: dependency as desideratum. Public Choice 124:19–31
Bünger B, Prinz A (2010) Staatliche Glücksförderung? Karl Popper, Richard Layard und das Rauchen. Ordo Jahrbuch für die Ordnung von Wirtschaft und Gesellschaft 61:169–190
Cigno A (2011) The economics of marriage. Perspektiven der Wirtschaftspolitik 12 (Issue Supplements1):28–41
Clark AE, Oswald AJ (1994) Unhappiness and unemployment. Econ J 104:648–659
Clark AE, Oswald AJ (2002) A simple statistical method for measuring how life events affect happiness. Int J Epidemiol 31(6):1139–1144
Conway E (2009) 50 economics ideas you really need to know. Quercus Publishing Plc, London
De Prycker V (2010) Happiness on the political agenda? PROS and CONS. J Happiness Stud 11(5):585–603
Diener E (2006) Guidelines for national indicators of subjective well-being and ill-being. Appl Res Qual Life 1:151–157
Diener E, Lucas RE, Schimmack U, Helliwell JF (2009) Well-being for public policy. Oxford University Press, Oxford
Di Tella R, MacCulloch R (2005) Gross national happiness as an answer to the Easterlin paradox? Available at: http://www.people.hbs.edu/rditella/papers/WPGNHappiness.pdf. Accessed on 19 Jan 2012
Di Tella R, MacCulloch R, Oswald A (2001) Preferences over inflation and unemployment: evidence from surveys of happiness. Am Econ Rev 91(1):335–341
Duden (2009) Wirtschaft von A bis Z. Bibliographisches Institut, Mannheim
Dulleck U, Kerschbamer R (2006) On doctors, mechanics, and computer specialists: the economics of credence goods. J Econ Lit 44:5–42
Duncan G (2010) Should happiness-maximization be the goal of government? J Happiness Stud 11(2):163–178
Epicurus (341–270 BC) Principle doctrines. (trans: Robert Drew Hicks 1925). Available at: http://classics.mit.edu/Epicurus/princdoc.html. Accessed on 5 Aug 2011
Flaschel P, Greiner A, Luchtenberg S (2010) Neo-Schumpeterian flexicurity capitalism: heterogeneous skills and the formation of elites. Homo Oeconomicus 27(1/2):25–67
Fors F (2010) Happiness in the extensive welfare state: Sweden in a comparative European perspective. In: Greve B (ed) Happiness and social policy in Europe. Edgar Elgar, UK, USA, pp 120–135
Forum Flexicurity (2008) Flexicurity—a European approach to labour market policy (with contributions of S. Bekker, T. Wilthagen, P.K. Madsen, J. Zhou, R. Rogowski, M. Keune and A. Tangian). Intereconomics 43:68–111
Frey BS, Stutzer A (2000) Maximizing happiness? German Econ Rev 1(2):145–167
Frey BS, Stutzer A (2005) Beyond outcomes: measuring procedural utility. Oxford Econ Pap 57(1):90–111
Frey BS, Stutzer A (2010) Happiness and public choice. Public Choice 144(3–4):557–573
Frey BS, Benz M, Stutzer A (2004) Introducing procedural utility: not only what, but also how matters. J Theo Inst Econ 160:377–401
Funk L (2008) European flexicurity policies: a critical assessment. Int J Comp Labour Law Ind Relat 24(3):349–384
Greve B (2010) Happiness and social policy in Denmark In: Greve B (ed) Happiness and social policy in Europe. Edgar Elgar, UK, USA, pp 136–145
Gul F, Pesendorfer W (2007) Welfare without happiness. Am Econ Rev 97(2):471–476
Jones R, Whitehead M, Pykett J (2009) From the nanny state to the avuncular state? Available at: http://governingtemptation.wordpress.com/2009/06/hello-world/. Accessed on 18 July 2011

Kahneman D, Krueger AB, Schkade D, Schwarz N, Stone A (2004) Toward national well-being accounts. Am Econ Rev 94(2):429–434
Kirchler E (2007) The economic psychology of tax behaviour. Cambridge University Press, Cambridge
Laibson D (1997) Golden eggs and hyperbolic discounting. Q J Econ 112:443–477
Layard R (1980) Human satisfaction and public policy. Econ J 90(360):737–750
Layard R (2005) Happiness: lessons from a new science. Penguin, New York
McChesney FS (1997) Money for nothing: Politicians, rent extraction and political extortion. Harvard University Press, Cambridge
McMahon D (2004) From the happiness of virtue to the virtue of happiness: 400 B.C.—A.D. 1780. Daedalus 133:5–17
McMahon D (2006) The pursuit of happiness—a history from the Greeks to the present. Penguin Books, London et al
Ng YK (2008) Happiness studies: Ways to improve comparability and some public policy implications. Econ Rec 84(265):253–266
Ng Y-K, Ho LS (2006) Introduction: happiness as the only ultimate objective of public policy. In: Ng Y-K, Ho LS (eds) Happiness and public policy: theory, case studies and implications. Palgrave Macmillan, Houndmills, pp 1–16
OECD (2011) Measuring well-being and progress. OECD, Paris
Ott J (2010) Greater happiness for a greater number: Some non-controversial options for governments. J Happiness Stud 11(5):631–647
Ouweneel P (2002) Social security and well-being of the unemployed in 42 nations. J Happiness Stud 3(2):167–192
Pacek AC, Radcliff B (2008a) Assessing the welfare state. The politics of happiness. Perspect Politics 6:267–277
Pacek AC, Radcliff B (2008b) Welfare policy and subjective well-being across nations: An individual-level assessment. Soc Indic Res 89(1):179–191
Plato (about 380 BC) The republic (trans: Benjamin Jowett). Available at: http://www.filepedia.org/files/Plato%20-%20The%20Republic.pdf. Accessed on 5 Aug 2011
Popper KR (1974) The Open Society and its enemies, Vol. 2: The high tide of prophecy: Hegel, Marx, and the aftermath. Routledge, London
Prinz A (2009) Accounting for well-being. In: Prinz A, Hospers G-J, Bünger B (eds) Squaring the circle: essays in honour of Bert Steenge. Lit-Verlag, Berlin, pp 88–112
Prinz A, Bünger B (2011a) Subjective well-being and the political pursuit of happiness: a Popperian view. Discussion Paper, University of Münster
Prinz A, Bünger B (2011b) The usefulness of a happy income index. Discussion Paper, University of Münster
Prinz A, Bünger B (2012) Balancing 'full life': an economic approach to the route to happiness. J Econ Psychol 33:58–70
Schwartz B (2004) The tyranny of choice. Sci Am 290:70–76
Schwartz B, Ward A, Monterosso J, Lyubomirski S, White K, Lehman D (2002) Maximizing versus satisficing: happiness is a matter of choice. J Pers Soc Psychol 83(5):1178–1197
Schwarz N (1999) Self-reports: how the questions shape the answers. Am Psychol 54:93–105
Schwarz N (2003) Self-reports in consumer research: the challenge of comparing cohorts and cultures. J Consum Res 29:588–594
Schwarz N (2007) Retrospective and concurrent self-reports: the rationale for real-time data capture. In: Stone A, Shiffman SS, Atienza A, Nebeling L (eds) The science of real-time data capture: self-reports in health research. Oxford University Press, New York, pp 11–26
Sparshott F (1994) Taking life seriously. A study of the argument of the 'Nicomachean Ethics'. University of Toronto Press, Toronto
Sunstein CR, Thaler RH (2003) Libertarian paternalism is not an oxymoron. Univ Chicago Law Rev 70:1159–1202
Sunstein CR, Thaler RH (2006) Preferences, paternalism, and liberty. In: Olsaretti S (ed) Preferences and well-being. Cambridge University Press, Cambridge, pp 233–264

Thaler RH, Sunstein CR (2003) Libertarian paternalism. Am Econ Rev 93(2):175–179
Thaler RH, Sunstein CR (2008) Nudge—improving decisions about health, wealth, and happiness. Yale University Press, New Haven
The New Paternalism (2006) The avuncular state. A smarter, softer kind of paternalism is coming into style. Economist 379:75–77
van Praag B, Ferrer-i-Carbonell A (2004) Happiness quantified: a satisfaction calculus approach. Oxford University Press, Oxford
van Suntum U (1991) Die soziale Komponente in der Marktwirtschaft. Aus Politik und Zeitgeschichte 41:16–27
Veenhoven R (2000) Well-being in the welfare state: level not higher, distribution not more equitable. J Comp Policy Anal Res Pract 2:91–125
Veenhoven R (2004) Happiness as a public policy aim: the greatest happiness principle. In: Linley PA, Joseph S (eds) Positive psychology in practice. Wiley, Hoboken, pp 658–678
Verme P (2007) Happiness and freedom. Working Paper, University of Torino and School of Management, Milano
Viebrock E, Clasen J (2009) Flexicurity and welfare reform: a review. Socio-Econ Rev 7(2):305–331
Winkelmann L, Winkelmann R (1995) Happiness and unemployment: a panel data analysis for Germany. Appl Econ Q 41(4):293–307
Winkelmann L, Winkelmann R (1998) Why are the unemployed so unhappy? Evidence from panel data. Economica 65(257):1–15
Wolinsky A (1995) Competition in markets for credence goods. J Inst Theor Econ 151:117–131

Chapter 14
A "Happiness Test" for the New Measures of National Well-Being: How Much Better than GDP are They?

Jan Delhey and Christian Kroll

14.1 Introduction

There is currently a broad global movement away from considerations of mere economic success towards a new public policy goal involving a broader notion of quality of life. This movement has also spurred the rethinking of which statistics inform us best about a country's situation and how its citizens are faring. For decades, the gold standard was a macroeconomic indicator: the GDP—gross domestic product, calculated per capita.[1] This is the most prominent yardstick that the media, politicians and the public consider when they try to assess how a country is performing. However, this measure was never meant to be a measure of the welfare of nations (as its creator Simon Kuznets already warned in the 1930s) and so there is growing skepticism about the GDP's usefulness as a measure of national well-being. Slogans such as "beyond GDP" or "redefining progress" challenge the preoccupation with the GDP. Back in the 1960s, Robert Kennedy expressed his uneasiness as follows:

---

[1] The gross domestic product (GDP), gross national product (GNP) and gross national income (GNI) are all measures of national economic output that are used interchangeably in this chapter. The GDP is the total value of the final goods and services produced within a country's borders in a year. The GNP is the total value of all the final goods and services produced by a country's factors of production and sold on the market in a year. Last but not least, the GNI comprises the total value of the goods and services produced within a country, together with its income received from other countries (notably interest and dividends), less similar payments made to other countries (Black 2003).

J. Delhey (✉)
School of Humanities and Social Sciences, Jacobs University Bremen, Bremen, Germany
e-mail: j.delhey@jacobs-university.de

C. Kroll
Bremen International Graduate School of Social Sciences, Bremen, Germany
e-mail: c.kroll@jacobs-university.de

H. Brockmann and J. Delhey (eds.), *Human Happiness and the Pursuit of Maximization*, Happiness Studies Book Series, DOI: 10.1007/978-94-007-6609-9_14,

> The Gross national product [GNP] does not allow for the health of our children, the quality of their education or the joy of their play. It does not include the beauty of our poetry, or the strength of our marriages, the intelligence of our public debates or the integrity of our public officials. It measures neither our wit nor our courage, neither our wisdom nor our devotion to our country. It measures everything, in short, except what makes life worthwhile (Senator Robert Kennedy in a speech at the University of Kansas on March 18, 1968).

Since then, increasing numbers of social scientists, politicians, and ordinary citizens have begun to adopt this view. To meet the new demand, several alternative national performance measures have been developed, which either try to heal some of the conceptual problems of the GDP, complement the GDP with other indicators, or replace the GDP entirely. But do these new measures indeed make a better job of capturing "what makes life worthwhile"? This is our guiding question in this book chapter, which we address from the perspective of human happiness—the quality of life (QOL) as experienced by ordinary citizens (Veenhoven 2007). In other words: How well do the new measures of well-being perform compared with the GDP when seen through a "happiness lens"? Are those new measures of national well-being better able to capture what makes people happy and satisfied with their lives?

The chapter is organized as follows: we will start by briefly reviewing the rise and fall of the GDP as a welfare measure; then, some alternatives are sketched out; finally, we turn to data on subjective well-being from 34 OECD countries to ascertain whether the GDP indeed performs badly in predicting a population's overall happiness and whether alternative measures perform better. The surprising answer is that from the happiness perspective, there is little wrong with the GDP, and most alternative QOL measures do not outperform the GDP. Yet, one measure does do a better job, and we close with the lessons happiness research provides for the construction of new—and better—welfare measures.

14.2 GDP/GNI: Their Rise and Fall

When the GDP was invented in the 1930s, it was not meant for metering a country's overall well-being. Rather, it was intended for obtaining an idea of the direction in which the economy was moving, which was a pressing issue in the Great Depression of the 1930s. For that purpose, the GDP "compresses the immensity of a national economy into a single data point of surpassing density" (Gertner 2010, 1). More precisely, it adds up all the goods and services produced in an economy within a year. Thus, it has the advantage of being able to total entities with different units and to summarize them in one single monetary figure. Moreover, once the figure has been adjusted per capita and purchasing power parity, it can be easily compared across nations. Last but not least, the assumption behind using the GDP to assess well-being is that the higher the level of economic production, the better people are able to satisfy their needs. Arguably, in the post-war decades this made perfect sense, as people's main concern was to improve their economic living situation. As a matter of fact, the material conditions in the industrialized world since 1945 have improved tremendously and it is right to assume that this has led to a significant increase in people's quality of life.

However, as the Stiglitz Commission pointed out in the latest comprehensive survey of the weaknesses of the GDP (Stiglitz et al. 2009), the figure has a number of important downsides that should remind politicians to be cautious. First, the GDP does not take into account the distribution of income. The mean per capita does not contain any information on whether this reflects how the money is actually distributed among the citizens of a country. Second, we do not know from looking at the GDP whether the money is really spent on improving people's living conditions. Third, the GDP does not account for sustainability and informal labor. In other words, current wealth may be produced at the expense of future generations, and a large part of production does not even feature in the GDP figure. In fact, a number of factors that are important for well-being (e.g., civic engagement, leisure, helping neighbors, informal childcare) are not captured because they take place outside the market. All of these caveats have led to numerous initiatives to find a better measure of the well-being of nations, which will be portrayed and then assessed from a happiness perspective in the remainder of this chapter.

14.3 The Search for a Better Measure: Three Main Strategies

Three key strategies have been employed to develop a better measure of well-being: *healing* the GDP, *complementing* the GDP, and *replacing* the GDP; we will briefly discuss each of these strategies. There is insufficient room here to undertake an extensive review of all the measures that have been suggested recently (Booysen 2002; Commission on the Measurement of Economic Performance and Social Progress 2009; Hagerty et al. 2001; Kroll 2011). Instead, our main goal is to explain the logic behind the three major kinds of strategies.

Healing the GDP

The first group of initiatives tries to deal with the aforementioned downsides of the GDP by attempting to fix the indicator itself. In other words, the mode of calculation is adjusted in a way that hopes to overcome the existing weaknesses without departing too much from the basic logic of a measure that seeks to sum up goods and services in a comparable, monetary figure. The examples listed in the respective column in Fig. 14.1 function according to the same logic as the GDP in that they contain quantified information about different entities expressed as a

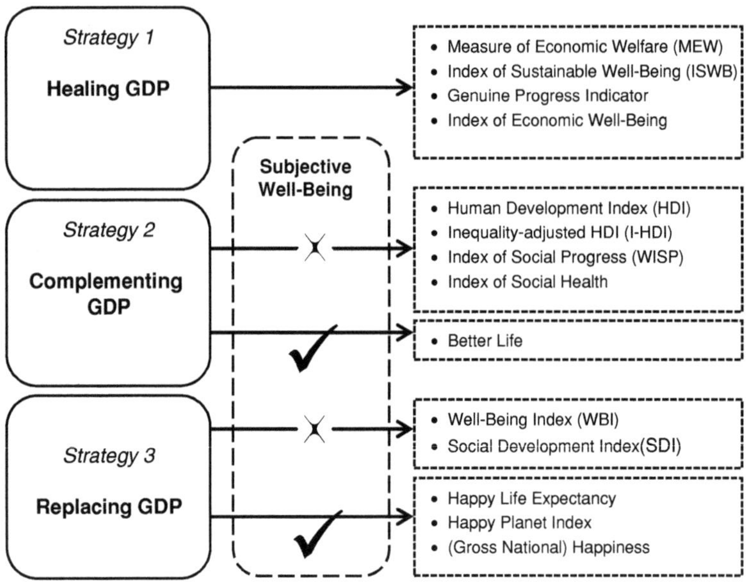

Fig. 14.1 Strategies for moving "beyond GDP" and examples. *Note* X = without subjective well-being measures; ✔= with subjective well-being measures

monetary value. In contrast to the GDP, though, they are enriched with mainly social and environmental factors in order to address the blind spots of the GDP. The latter components are first monetized and then added to or subtracted from the original value of the GDP.

One key aim of this group of measures is to account for sustainability and the environmental damage associated with GDP growth. For example, the Index of Sustainable Economic Welfare (Cobb and Cobb 1994) and the Genuine Progress Indicator (Cobb et al. 1999) are both based on the consumption of private households. However, they also reflect additional social factors such as household labor and education with a rising value, while air pollution and environmental damage lower the score. As a consequence, the downsides of economic growth and modernization ought to be accounted for whilst retaining the benefits of the GDP, namely a single figure that captures different entities and is comparable across nations. In practice, however, the monetization of social and environmental factors contains a number of value judgments by the researcher and is therefore controversial. On balance, these kinds of well-being measure provide a step in the right direction as well as an important tool to engage with the downsides of the GDP for anyone keen to retain the advantages of the indicator.

Complementing the GDP

The second group of measures moves further away from the GDP as a yardstick than the previous approaches but does not abandon the sum of goods and services altogether. Instead, this group of measures seeks to assess national well-being by complementing the GDP with a number of key social indicators. In contrast to the aforementioned group, though, social and environmental factors are usually not forced into the logic of the GDP by means of monetization. Instead, the other indicators are standardized and subsequently merged with the GDP into a new index comprising a number of dimensions.

For example, the Human Development Index (see also the more detailed portrayal below) comprises the three dimensions health, education, and material living conditions, which are measured by life expectancy, years of schooling, and GNI, respectively. The three dimensions are standardized in order to produce a sub-index for each dimension that is then summarized into the total HDI score through a method of equal weights (one-third for each of the three dimensions).

While the method of complementing the GDP with further indicators is able to overcome the controversial monetization from which the measures that try to "heal GDP" suffer, the standardization of different units is also controversial. In particular, merging different units into a single standardized index is methodologically challenging and again requires value judgments by the researcher. Furthermore, it is unclear which precise dimensions ought to be included in such an index and how many are sufficient to deliver a good picture of the quality of life in a country. Nonetheless, this group currently provides the most prominent and widely accepted measures of national well-being in the sense of a safe middle ground between the two other approaches outlined here.

Replacing the GDP

The most radical departure from the GDP is embodied by the third group of measures, which seeks alternative indicators of well-being without accounting for the sum of goods and services produced in an economy. The logic behind this approach is that the GDP has always been and remains a means to an end rather than the end itself. Thus, according to this approach it would be more appropriate to examine the key indicators that really make life worth living rather than looking at economic production, which merely serves to improve those other indicators.

Famous examples include the Happy Planet Index (nef 2009) calculated by the New Economics Foundation (nef). The index comprises life expectancy, life satisfaction, and the ecological footprint and is therefore able to demonstrate how many resources countries need in order to produce a certain level of health and subjective well-being (SWB). As a result, Latin American countries, which have high levels of SWB despite smaller ecological footprints, top the list (with Costa

Rica in the lead). In the end, industrialized Western nations are dragged down compared with their traditional GDP rankings as they require far more resources to produce comparable levels of health and life satisfaction.

Another example is the Happy-Life-Expectancy (Veenhoven 1996, 2005), which merges life expectancy and people's subjective satisfaction with life into one single index. Yet another approach is to eliminate objective indicators entirely and to rely exclusively on people's own evaluation of their quality of life. This approach, in reference to the GDP also known as Gross National Happiness (Veenhoven 2007), argues that citizens are best equipped to evaluate their lives, and that relevant life circumstances translate into more or less positive evaluations, after being filtered by personal life goals, aspirations, and social comparisons.

Replacing the GDP altogether is quite a drastic strategy for assessing national well-being, as not only is economic growth a prerequisite for many of the social goods that make life enjoyable but the metric of GDP is also highly correlated with such other factors (Kassenböhmer and Schmidt 2011). Thus, by arguing that the GDP is only a means to an end, these measures are in danger of making a conceptual assumption that is noble in theory but can be challenged in practice on the basis of actual causal mechanisms and empirical data.

14.4 Six Alternative Measures Portrayed

In this section we briefly portray six widely used QOL measures, which we later put in a horse race against income measures (GNI). Since this horse race is run for 34 OECD countries, we selected measures that are available for a broad number of countries, which excludes "strategy 1 measures" (healing the GDP). Since we will use information on citizens' subjective well-being to judge how well the GNI and other measures "measure what makes life worthwhile" (Kennedy), we further exclude from both the "strategy 2 measures" and the "strategy 3 measures" those that rely in part or entirely on information on subjective well-being. The exception is the "OECD Better Life Index," which we turned into a measure without SWB by excluding the respective dimension. This leaves us with six measures, four of them complementing the GDP and two replacing the GDP (see Fig. 14.1—the measures we portray are printed in bold).

Human Development Index

A long-standing and perhaps the most prominent alternative measure is the Human Development Index (HDI), which is based on Sen's capabilities approach (Sen 1993). The HDI integrates health, education, and economic affluence into a human development framework, and it is this pluralistic conception that Sen himself regards as groundbreaking (Sen 2000). More precisely, five indicators are merged in a two-step process: first into three domain indices and then into an overall HDI score, which can take values between 0 and 1. Although some technicalities of the index construction have been repeatedly criticized (Lind 2004, 2010), the HDI has received a great deal of attention and is currently the main rival of the GDP. Almost all the OECD countries are ranked in the category "very high human development," with Norway, Australia, and New Zealand leading the HDI league table (UNDP 2010). Turkey (lowest), Mexico, and Chile are the OECD laggards, but still rank as having "high human development."

Inequality-Adjusted Human Development Index

In the 2010 Human Development Report (UNDP 2010), for the first time an inequality-adjusted HDI is presented: the I-HDI. This innovation reflects the criticism that the statistical means from which the HDI scores are calculated do not contain any information on how health, income, and education are distributed across the population. How citizens fare crucially depends on the distribution of human development within a country, though. To heal this conceptual problem, each of the three human development components account for distribution issues. Consequently, the overall index is inequality-adjusted as inequalities are integrated in such a way that the percentage loss in human development is calculated. In other words, the I-HDI tells us how much the human development in a country falls short of the potential human development under the condition of complete equality. Logically, the I-HDI values are always lower than the HDI values. Among the OECD countries, the most developed are Norway, Australia, and Sweden[2]; the laggards are Turkey (lowest), Chile, and Mexico.

OECD Better Life Index

The recently launched *Better Life Initiative* (OECD 2011) features a compendium of well-being indicators, as well as a new composite index. The Better Life framework distinguishes between two main concepts: material living conditions and quality of life. Material living conditions are metered in three life domains: income and wealth; jobs and earning; and housing. Quality of life is measured in eight life domains: health status; work and life balance; education and skills; civic engagement and governance; social connections; environmental quality; personal security; and subjective well-being. With these 11 life domains, the OECD aims to examine the most relevant features that shape people's lives. In total, twenty-one social indicators are utilized for quantifying these life domains. The OECD

[2] We have no I-HDI data for New Zealand, which ranks third in the HDI.

compendium reports country rankings across all 21 indicators, but does not deliver an authoritative index. The idea behind this strategy is to show the complexity of well-being, rather than brushing over this complexity by condensing everything into one single number. However, the OECD project website provides an easy tool for computing such an index, whereby the user is invited to apply his or her preferred weighting to the 11 life domains. In this chapter we will use an index in which every life domain carries equal weight, bar subjective well-being, which is excluded. The simple reason is that in the next section we will explain the international variation in subjective well-being by national performance measures, hence SWB cannot appear on both sides of the equation. In this equal-weight index (ex. SWB), the countries scoring highest on the OECD index are Australia, Canada, and Sweden/New Zealand, which are on a par; the countries scoring lowest are Turkey (lowest), Mexico, and Chile. No matter how the domains are weighted, the theoretical range of the Better Life Index is between 0 and 10, and the higher the index value, the better is life.

Index of Social Progress

The *Index of Social Progress* (Estes 1998, 2010) is probably one of the most encompassing national performance measures. Its main conceptual focus is social development, more precisely adequacy of social provision. This concept refers to "the changing capacity of governments to provide for the basic social, material, and other needs of the people living within their borders, e.g., for food, clothing, shelter, and access to at least basic health, education, and social services" (Estes 1998). In total 41 indicators are used to capture nine dimensions: education, health status, women's status, defense effort (-), economy, demography, environment, social chaos (-), and cultural diversity; (-) denotes domains that are, conceptually, negatively linked to social progress. Separate factor analyses are run over the nine domains to create a subindex score for each domain. These subindices are then further factor analyzed in order to arrive at the WISP, the Weighted Index of Social Progress (cf. Estes 2010, p. 367 and Table 2 for a more detailed description of this procedure). The WISP is calculated worldwide and allows comparisons over time. Among the OECD countries, Sweden, Denmark, and Norway are the "social leaders," whereas Turkey (lowest), Mexico and South Korea are the "social laggards."

Well-Being Index

The Well-Being Index (McGillivray 2005) is a close cousin of the Human Development Index. In fact, it uses exactly the same indicators, except per capita income, which is excluded altogether. Hence, the WBI merges information from two domains, health and education, only. The second difference is *how* the data are merged. Whereas the HDI uses a predefined weighting scheme, the WBI uses factor analysis (principal component analysis): it is the strongest factor that emerges out of life expectancy, adult literacy, and gross enrollment. The OECD countries with the highest well-being are Australia, Belgium, and the United Kingdom; those with the lowest well-being are Turkey (lowest), Mexico, and Hungary. For our purpose, the WBI is a valuable indicator exactly because it omits the economic dimension entirely.

Social Development Index

The Social Development Index (SDI) was originally constructed in 1989 and updated in 2008 with newer cross-national data (Ray 2008). The composite index follows on from the Physical Quality of Life Index (PQLI) but claims to be a better tool for international comparisons as it is a broader, multi-dimensional approach. The SDI indeed includes 10 development indicators (rather than 3 as in the PQLI) to represent social development across 102 countries. The components are life expectancy, adult literacy rate, gross enrollment ratio, infant survival rate, supply of calories, proteins, and fat per day, respectively, telephone lines per 1,000 people, physicians per 100,000 people, as well as electricity consumption. The ranking is topped by Norway, Sweden, and the US, while Mexico (lowest), Chile, and Turkey are at the bottom of the SDI.

14.5 Do the New Measures Outperform the GDP? Some Empirical Insights from the Perspective of Human Happiness

As we have seen, the GDP has mainly attracted criticism as a measure of societal well-being because of its narrow focus on economic production. In fact, all theoretical conceptions of individual quality of life emphasize the multi-dimensionality of what matters to people (for a review, see Philips 2006). Erik Allardt, for instance, defined the three pillars of personal quality of life as having, loving, and being (Allardt 1993). In a similar fashion, Maslow (1943) claimed earlier that humans are motivated by five basic needs, which are hierarchically ordered: physiological needs; safety needs; social needs; esteem needs; and self-actualization needs. Doyal and Gough (1991) produced a more extensive list of 11 intermediate needs, covering, among other things, housing, health care, relations with others, economic security, and education.

Bearing these conceptions in mind, it is more than plausible that broader QOL measures should capture what matters to people better than the GDP. This expectation can also be derived from *individual*-level happiness research, which has repeatedly demonstrated that a number of things make individuals happy—among them income and a comfortable living standard, but many other things as well (Layard 2005). Layard lists the "big seven" individual-level factors affecting happiness as family relationships, financial situation, work, community and friends, health, personal freedom, and personal values. Likewise, when Europeans are asked what matters for their personal quality of life, almost everywhere people pick the triad of income, family, and health (Delhey 2004).

The following empirical exercise investigates whether, as measures of *national* well-being, the new QOL measures outperform the GDP. Our yardstick is *average* self-reported happiness with life as a whole (Veenhoven 1984). Arguably, the things that really matter to people (remember Kennedy's verdict) should show up on their personal balance sheet of life—their overall happiness. If the new QOL measures are better able to capture these salient concerns than the GDP, the former should be better at predicting average happiness than the latter.

Measuring Happiness

For the concept of happiness as defined above, three different concrete measures are common in cross-national research (Veenhoven 2007):

Cantrils's ladder of life, also called life contentment (example from Gallup World Poll):

> Please imagine a ladder with steps numbered from zero at the bottom to 10 at the top. The top of the ladder represents the best possible life for you and the bottom of the ladder represents the worst possible life for you. On which step of the ladder would you say you stand at this time?

Life satisfaction (example from the World Values Survey):

> All things considered, how satisfied are you with your life as a whole these days? Please use this card to help with your answer [1 dissatisfied (…) 10 satisfied].

Life happiness (example from the World Values Survey):

> Taking all together, how happy would you say you are: very happy, quite happy, not very happy, not at all happy?

We collected data on these three measures for 34 OECD countries. The contentment scores came from the Gallup World Poll 2010, reported by the OECD (2011). The life satisfaction and happiness scores were computed from either the World Values Surveys or the European Values Study 2008; for each country, we took the latest year available. At the country level, all three measures are highly correlated: contentment–life satisfaction at 0.77; contentment–happiness at 0.69; and life satisfaction–happiness at 0.81. We merged all three into one single score of SWB.[3] The resulting relative scores meaningfully differentiate the degree of life enjoyment; the higher the score, the higher a population's overall subjective well-being. As Fig. 14.2 shows, Denmark, the Netherlands, and Norway are the countries where citizens—on average—enjoy life most, whereas in Hungary, Estonia, and Turkey, life enjoyment is the lowest.

Some Simple Correlations

How well does the GDP predict the national SWB? As Fig. 14.3 shows, the correlation between GNI per capita (in purchasing power parities, to adjust for different price levels) and average SWB is very strong (0.58) and highly significant. By and large, people enjoy life more in richer countries. There is also no sign of a leveling of this association among the richest nations, as it is often claimed (Inglehart and Klingemann 2000). If one uses a logged scale of GNI per capita, as advocated by many economists, the regression line is even slightly concave, again suggesting that there is no decreasing marginal utility of national income.

Do the new QOL measures perform better than the GNI? The surprising answer is no, not across the board (see Table 14.1). Only one single measure outperforms the GNI, namely the Better Life Index (exclusive of subjective well-being). The correlation coefficient between Better Life and SWB is a highly significant 0.67. In contrast, most other measures, from HDI over WBI and I-HDI to SDI, perform slightly worse than the GDP, with coefficients between 0.51 and 0.47. Finally, the Index of Social Progress (WISP) performs considerably worse than the GDP. The WISP's mild correlation with the average SWB does not reach the usual threshold of statistical significance of $p < 0.05$. This is somewhat surprising, given that the WISP is a convergence of so many social indicators, yet it suggests that the normative assumptions underlying the WISP of what social progress entails are not fully shared by ordinary citizens. Obviously, the WISP includes indicators that are irrelevant to human happiness—or if relevant, it treats them in a way that does not conform to the OECD citizens' happiness calculus.[4] Figure 14.4 displays the scatterplots for the best- and the worst-performing QOL measures, the Better Life Index and the Index of Social Progress.

These findings are confirmed if we look at the three component parts of SWB separately (Table 14.1). Across the three components, the Better Life Index is most closely associated with people's enjoyment of life, followed by the GNI. However, Better Life typically wins by a narrow margin only. Equally consistently, the WISP has the weakest association with all three SWB components, and

[3] The measures were first transformed into Z-scores and then added up. Each SWB component was assigned equal weight.

[4] For example, low fertility enters the WISP with a negative sign—as social regress. Yet for many people, having few children indicates control over one's life, which is valued by many humans.

Fig. 14.2 SWB in OECD countries

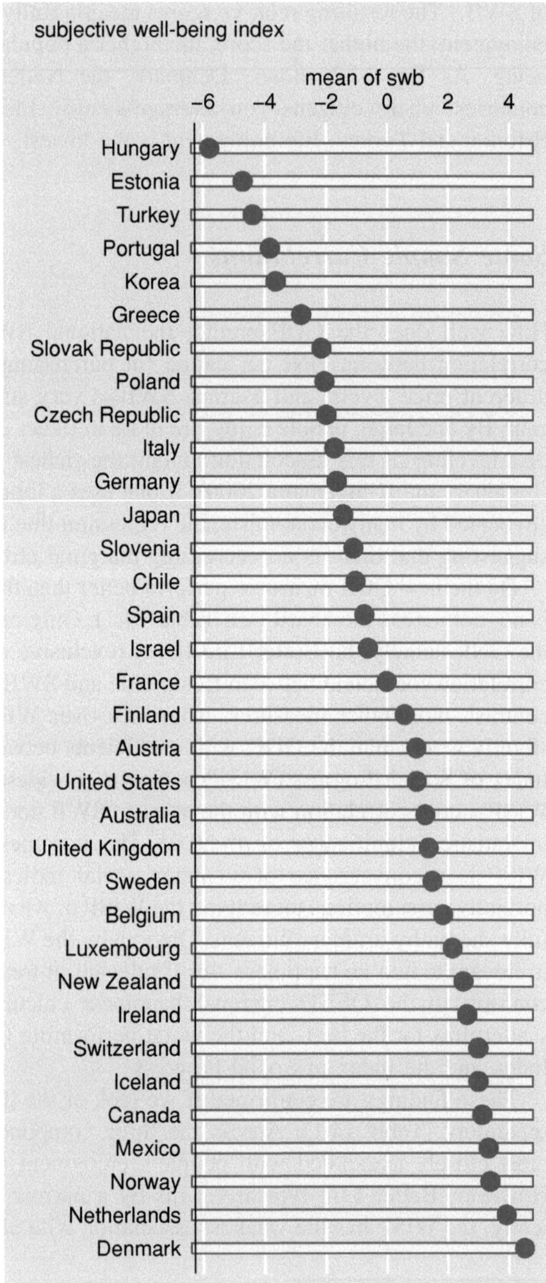

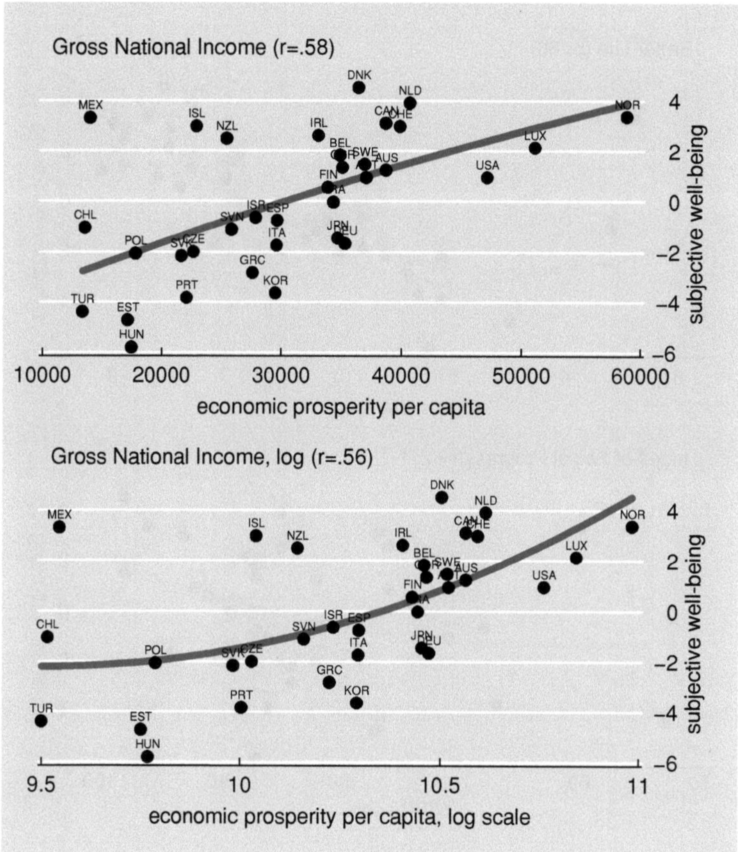

Fig. 14.3 Association between national income and SWB

Table 14.1 Correlations of composite QOL measures and average SWB

	SWB index	Life contentment	Life satisfaction	Feeling happy
Better life	0.66***	0.71***	0.56***	0.54***
GNI	0.58***	0.67***	0.46**	0.47**
HDI	0.51**	0.63***	0.37*	0.41*
WBI	0.51**	0.57***	0.36*	0.45**
I-HDI	0.48**	0.56***	0.41*	0.34^{+}
SDI	0.47**	0.57***	0.38*	0.34^{+}
WISP	0.27	0.24	0.35*	0.16

$^{+}$ $p < 0.10$; * $p < 0.05$; ** $p < 0.01$; *** $p < 0.001$

reaches the typical threshold of statistical significance for just one component, life satisfaction. Another important message emerges from Table 14.1: of the three SWB components, the average life contentment is most closely associated with the

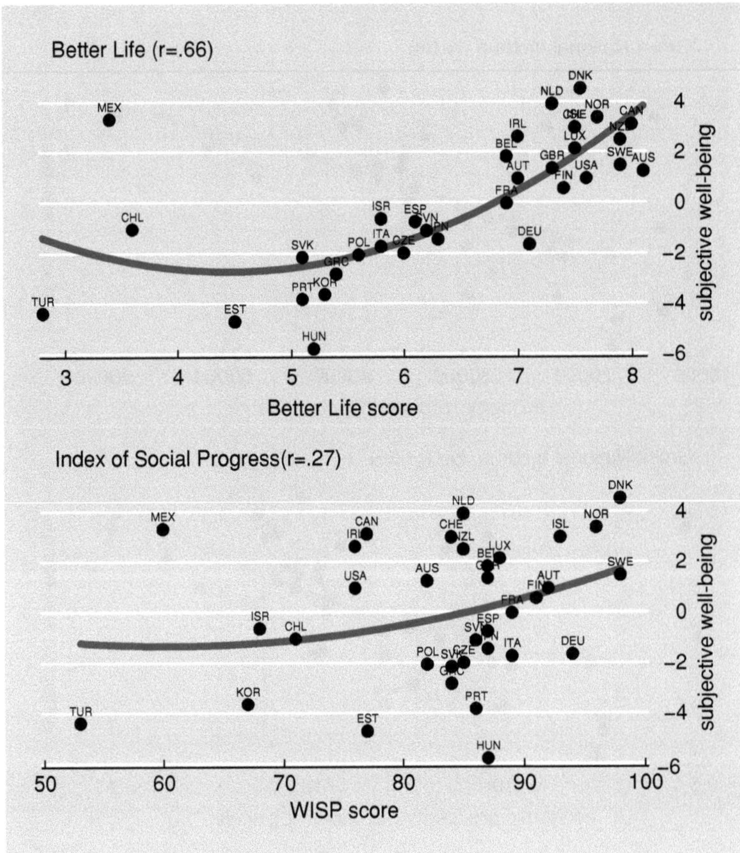

Fig. 14.4 Association between two QOL measures and SWB

objective well-being indexes, regardless of which one we consider. Across the board, the correlation coefficients are higher for life contentment than for life satisfaction and happiness. This again proves that the three concrete measures of life enjoyment are not fully interchangeable, and that the results produced with one measure do not necessarily hold for the other measures as well (Bjornskov 2010).

Explaining National SWB

It is well known that overall life enjoyment cannot *fully* be explained by societal conditions, and this is suggested by our correlations as well. At least to some extent, national or world-regional peculiarities influence SWB levels over and above societal conditions, which we can capture in accounts of objective living conditions. It is widely known, for example, that Latin Americans are "happiness overachievers"—they are more upbeat than their societal conditions suggest (Diener et al. 2000). In contrast, Eastern Europeans are known to be "happiness underachievers"—they are typically less satisfied than their societal conditions suggest, remaining in the aftermath of the system collapse from 1989 to 1992 (Deaton 2008; Inglehart et al. 2008). To some extent, these peculiarities also surface in the SWB scores presented here (Figs. 14.2, 14.3): whereas in the scattergrams Chile and in particular Mexico are positioned above the fit lines (SWB overachievers), Estonia and Hungary are positioned below the fit lines (SWB underachievers). However, this latter pattern is much less clear for the other post-communist OECD countries, the Czech Republic, Poland, Slovenia, and Slovakia. The next step of the analysis accounts for these world-regional peculiarities. With linear regressions, the average SWB is explained by national well-being (the various measures are used in turn), while controlling for "Latin America" and "Eastern Europe." For that purpose, two dummy variables distinguish between Latin American and other countries, and between Eastern European and other countries, respectively. To make the size of the coefficients comparable across the various QOL measures, the latter are standardized into Z-scores, so that they no longer use a different metric. The regression results are shown in Table 14.2.

When the two region dummies are included, the GNI exerts a strong and statistically highly significant influence on the average SWB: people do rate their lives better in richer countries, independently of which income scale is used, linear or logarithmic. Together with the two region dummies—of which only "Latin America" is significant—the GNI explains roughly half of the international variance in SWB. Again, only the Better Life Index turns out to have an advantage over income: its regression coefficient is higher (although still within the confidence interval of the income measures' coefficients), and together with the region dummies, the Better Life Index can account for roughly three-quarters of the variance in SWB. In contrast, most other QOL measures perform more or less similarly to income, and the WISP performs worse.

A useful extension of the analysis is to look at the group of the richest OECD nations separately, since quality of life, rather than income alone, is often seen as the key concern in affluent societies. We use two different thresholds for defining "rich" societies, one derived from the World System Theory (organic core countries, cf. Babones 2005), the other being a GDP per capita threshold of $25,000. This leaves us with 19 "core countries" and 20 "rich countries." For these two groups of countries, Better Life shows by far the strongest association with SWB, even more strongly than for the entire group of 34 countries (see Table 14.3). In contrast, most other well-being measures, including the GDP, are less strongly and less significantly associated than for the full sample. For the core countries in particular, the associations are typically weaker. The exception is the inequality-adjusted HDI, which correlates more strongly with SWB among core/rich countries than among all 34 OECD nations .

Table 14.2 Predicting SWB: OLS regressions

	m1 b/t	m2 b/t	m3 b/t	m4 b/t	m5 b/t	m6 b/t	m7 b/t	m8 b/t
Latin Am	3.953*	5.333**	3.476	4.871**	2.847	5.856**	2.460	6.345***
	(2.27)	(2.87)	(1.94)	(2.84)	(1.58)	(2.82)	(1.31)	(4.68)
Postcom	−1.259	−0.867	−2.075	−2.751**	−2.072	−1.214	−3.462**	−1.168
	(−1.14)	(−0.78)	(−1.95)	(−3.0 7)	(−1.84)	(−1.09)	(−3.34)	(−1.55)
GNI	1.765***							
	(3.81)							
GNI log		2.040***						
		(4.14)						
HDI			1.492**					
			(3.31)					
I-HDI				1.876***				
				(4.45)				
WBI					1.328**			
					(2.86)			
SDI						1.977**		
						(3.64)		
WISP							1.026*	
							(2.31)	
Better life								2.460***
								(7.27)
_cons	−0.010	−0.161	0.162	0.179	0.198	−0.299	0.466	−0.167
	(−0.02)	(−0.37)	(0.37)	(0.44)	(0.44)	(−0.64)	(1.03)	(−0.54)
r2	0.501	0.530	0.458	0.571	0.419	0.492	0.372	0.732
N	34	34	34	32	34	32	34	34

*$p < 0.05$; ** $p < 0.01$; *** $p < 0.001$
Latin Am. Latin American country; *Postcom.* Postcommunist country

Table 14.3 Correlations of QOL measures and average SWB, by country groups

	All 34 countries	Organic core countries only (19 countries)	Rich countries only (20 countries)
Better Life	0.66***	0.75***	0.81***
GNI	0.58***	0.38+	0.54**
HDI	0.51**	0.31	0.35+
WBI	0.51**	0.30	0.40+
I-HDI	0.48**	0.54*	0.65***
SDI	0.47**	0.32	0.51*
WISP	0.27	0.12	0.28

+ $p < 0.10$; *$p < 0.05$; **$p < 0.01$; ***$p < 0.001$

14.6 Conclusion: The Bumpy Road Towards a New Gold Standard

This chapter departed from Kennedy's famous dictum that the GNP "measures everything (...) except that which makes life worthwhile (...)." Following this line of thought, a movement that seeks to replace the GDP with a better, more encompassing summary indicator of well-being has gained momentum (see Kroll 2011 for an overview). Implicitly or explicitly, these new measures claim to capture better what makes life worthwhile. However, most new measures fail to deliver what they promise, if we use the average subjective well-being as expressed in representative surveys as the yardstick. Only the OECD's Better Life Index (to repeat: exclusive of SWB, of course) has an advantage over the GNI in this respect. In contrast, all the other QOL measures employed do not perform better in predicting subjective well-being than the GNI, and one measure—the Index of Social Progress—performs worse.

Several lessons can be drawn. One straightforward lesson is that economic activities and the affluence they create actually *do* make life worthwhile for a huge majority of people–even among the OECD countries, of which many are affluent. This suggests that what Samuelson and Nordhaus observed 60 years ago is still valid today: "People do not live by bread alone. Nor does society live by GNP alone. But on our way to that utopian state of affluence where concern about material well-being will disappear, we do need a summary measure of aggregate economic performance" (quoted from Sills and Merton 1991). Yet the question is whether it is affluence as such that nurtures life enjoyment in contemporary OECD societies, or the many good things and activities for which money can be used. In that sense, money can help satisfy a number of human needs, rather than the need to consume alone. Globally, national income correlates strongly with social progress in terms of health and education, and it also correlates with desired institutional qualities such as democracy and rule of law (Delhey and Newton 2005; Inglehart and Welzel 2005). In a similar fashion, data analyses for Germany have shown that much of the variation in alternative well-being indicators, as suggested by the Stiglitz Commission, can be captured well by economic indicators alone, especially the GDP and the unemployment rate (Kassenböhmer and Schmidt 2011). In short, modernization is a tight-knit syndrome, and prosperity is an integral part of it, as argued by the human empowerment theory (Inglehart and Welzel 2005).

Does this mean that the new performance measures are redundant? This conclusion would be premature–provided that the new measures are not too closely modeled after the concept of development/modernization. Although being broader than the GNI, the HDI, I-HDI, WBI, and SDI obviously do not capture *more* of the things that make life worthwhile, since they do not make a difference to human happiness. On the other hand, the example of the encompassing Index of Social Progress demonstrates that extreme broadness is not the trick either. Either the WISP's dashboard of indicators is flawed (in the sense that the researcher's idea of what constitutes progress deviates from the laymen's idea), or the mathematical index construction (e.g., too much weight is given to things that ordinary people do not value much, from a happiness perspective).

The measure that indeed *does* make a better job than the GNI of predicting national subjective well-being is the Better Life Index. This holds for all the OECD countries, and in particular for the subset of the richest/core countries. Obviously, its life domains and selection of indicators capture the life facets that OECD citizens truly value, over and above the achieved level of economic advancement. More research is needed here to find out *exactly what* these extra components are that give the Better Life Index added value. Given the fact that previous research has revealed that social capital (especially when compared with the GDP) matters more for the SWB of rich countries than for that of poorer ones (Kroll 2008), our informed guess is that issues of social capital, social cohesion, and greater equality make the difference here. The fact that the inequality-adjusted HDI, too, works very well for the rich/core countries points in a similar direction.

One idea for improving the Better Life Index further is to use measures of subjective well-being to calibrate the index. In our analysis, each of the ten Better Life domains carries equal weight, in the absence of a convincing theoretical argument for why the domains should be treated unevenly. Yet instead of assigning equal weights, the strength of the association between the various life domains and the average SWB could be used for assigning differential weights, so that the domains that matter greatly to people's life enjoyment enter the index with a stronger weight than those domains that matter less (see *The Economist*'s 2005, Quality-of-Life Index for a similar approach). Apart from the weighting issue, it might also turn out that there are better strategies than merging all the information into one single composite index. An alternative would be to factor analyze the Better Life data matrix in order to establish how many separable dimensions of national well-being are captured in the data. Yet another alternative would be to select from the dashboard of Better Life domains those that *together* explain the international variation in life enjoyment best, following the idea of "income + x."

To summarize our argument, for an economic indicator never intended to assess national well-being, the GDP is surprisingly successful in predicting a population's subjective well-being. At the same time, the theoretical claim of the social indicators movement about the multi-dimensionality of human concerns is a valid criticism, and conceptually it should be possible to come up with performance measures that embrace this multi-dimensionality better than an economic performance measure alone. Even if we lived in a world where the GDP performs exactly identically to a composite index of quality of life, on theoretical grounds one could still argue that the latter is more appropriate than the former for formulating evidence-based policy. Although a new gold standard in measuring national well-being has not been found yet, the Better Life Index demonstrates that progress towards this goal is possible. This chapter has demonstrated that a happiness perspective can add important insights along the way to facilitate the search for such a new, widely accepted, internationally comparable measure.

References

Allardt E (1993) Having, loving, being: an alternative to the Swedish model of welfare research. In: Nussbaum M, Sen A (eds) The quality of life. Clarendon Press, Oxford, pp 88–94

Babones S (2005) The country-level income structure of the world economy. J World Syst Res, XI, pp 29–55

Bjornskov C (2010) How comparable are the Gallup World Poll life satisfaction data? J Happiness Stud 11:41–60

Black J (2003) A dictionary of economics. Oxford University Press, Oxford

Booysen F (2002) An overview and evaluation of composite indices of development. Soc Indic Res 59:115–151

Cobb CW, Cobb JB (1994) The green national product: a proposed index of sustainable economic welfare. University of America, Lanham

Cobb C, Goodman GS, Wackernagel V (1999) Why bigger isn't better: the genuine progress indicator. Redefining Progress, San Francisco

Commission on the measurement of economic performance and social progress (2009) Survey on existing approaches to measuring socio-economic progress. Available at: www.stiglitz-sen-fitoussi.fr/documents. Accessed on 30 Apr 2012

Deaton A (2008) Income, health and wellbeing around the world: evidence from the Gallup World Poll. J Econ Perspect 22:53–72

Delhey J (2004) Life satisfaction in an enlarged Europe. European Foundation for the Improvement of Living and Working Conditions, Dublin

Delhey J, Newton K (2005) Predicting cross-national levels of social trust: global pattern or Nordic exceptionalism? Eur Sociol Rev 21:311–327

Diener E, Scollon CKN, Oishi S, Dzokoto V, Suh EM (2000) Positivity and the construction of life satisfaction judgements: global happiness is not the sum of its parts. J Happiness Stud 1:159–176

Doyal L, Gough I (1991) A theory of human need. Palgrave Macmillan, New York

Estes RJ (1998) Trends in world social development, 1970–1995. Development challenges for a new century. J Developing Soc 14:11–39

Estes RJ (2010) The world social situation: development challenges at the outset of a new century. Soc Indic Res 98:363–402

Gertner J (2010) The rise and fall of G.D.P. Times Magazine, New York

Hagerty MR, Cummins RA, Ferriss AL, Land K, Michalos A, Peterson M, Sharpe A, Sirgy J, Vogel J (2001) Quality of life indexes for national policy: review and agenda for research. Soc Indic Res 55:1–96

Inglehart R, Foa R, Peterson C, Welzel C (2008) Development, freedom, and rising happiness. A global perspective, 1981–2007. Perspect Psychol Sci 3:264–285

Inglehart R, Klingemann HD (2000) Genes, culture, democracy, and happiness. In: Diener E, Suh EM (eds) Subjective well-being across cultures. MIT Press, Cambridge, pp 165–183

Inglehart R, Welzel C (2005) Modernization, cultural change and democracy. The human development sequence. Cambridge University Press, New York

Kassenböhmer SC, Schmidt CM (2011) Beyond GDP and back: what is the value-added by additional components of welfare measurement? DIW: SOEP papers 351

Kroll C (2008) Social capital and the happiness of nations. The importance of trust and networks for life satisfaction in a cross-national perspective. Peter Lang Publishing, Frankfurt a.M

Kroll C (2011) Measuring progress and well-being: achievements and challenges of a new global movement. Friedrich Ebert Foundation, Berlin

Layard R (2005) Happiness. Lessons from a new science. Penguin Books, London

Lind N (2004) Values reflected in the human development index. Soc Indic Res 66:283–293

Lind N (2010) A calibrated index of human development. Soc Indic Res 98:301–319

Maslow AH (1943) A theory of human needs. Psychol Rev 50:296–370

McGillivray M (2005) Measuring non-economic well-being achievement. Rev Income Wealth 51:337–364
nef (2009) The happy planet index 2.0. The New Economics Foundation, London
OECD (2011) OECD better life initiative. Compendium of OECD well-being indicators. Organization of Economic Co-Operation and Development, France
Philips D (2006) Quality of life. Concept, policy and practice. Routledge, London
Ray AK (2008) Measurement of social development: an international comparison. Soc Indic Res 86:1–46
Sen A (1993) Capability and well-being. In: Nussbaum M, Sen A (eds) The quality of life. Clarendon Press, Oxford, pp 31–53
Sen A (2000) A decade of human development. J Hum Dev 1:17–23
Sills DL, Merton RK (1991) Social science quotations. Who said what, when, and where. McMillan Publishing Company, New York
Stiglitz JE, Sen A, Fitoussi J-P (2009) Report by the commission on the measurement of economic performance and social progress. Available at: http://www.stiglitzsen-fitoussi.fr/en/index.htm. Accessed on 30 Apr 2012
The Economist Intelligence Unit's quality-of-life index (2005) In: The Economist. The World in 2005, pp 1–4
UNDP (2010) Human development report 2010: the real wealth of nations: pathways to human development. United Nations Development Program, New York
Veenhoven R (1984) Conditions of happiness. Springer, Doordrecht
Veenhoven R (1996) Happy life expectancy: a comprehensive measure of quality of life in nations. Soc Indic Res 39:1–58
Veenhoven R (2005) Apparent quality-of-life in nations: how long and happy people live. Soc Indic Res 71:61–86
Veenhoven R (2007) Measures of gross national happiness. In: OECD (ed) Measuring and fostering the progress of societies. OECD Publishing, Paris, pp 232–253

Index

H. Brockmann and J. Delhey (eds.), *Human Happiness and the Pursuit of Maximization*, Happiness Studies Book Series, DOI: 10.1007/978-94-007-6609-9,

MIX
Papier aus verantwortungsvollen Quellen
Paper from responsible sources
FSC® C105338

If you have any concerns about our products,
you can contact us on
ProductSafety@springernature.com

In case Publisher is established outside the EU,
the EU authorized representative is:
Springer Nature Customer Service Center GmbH
Europaplatz 3, 69115 Heidelberg, Germany

Printed by Libri Plureos GmbH
in Hamburg, Germany